T0135556

CONTINUOUS CULTURE AND EXTRACELLULAR RECOMBINANT PROTEIN EXPRESSION IN *Escherichia coli*

Dissertation submitted for the award of the title of

Dr. rer. nat.

Fermentation Engineering

Faculty of Technology

Bielefeld University

25.06.2014

Ram Shankar Velur Selvamani

Born 14.06.1985 (Chennai, India)

Bibliografische Information der Deutschen Nationalbibliothek

Die Deutsche Nationalbibliothek verzeichnet diese Publikation in der
Deutschen Nationalbibliografie; detaillierte bibliografische Daten sind
im Internet über http://dnb.d-nb.de abrufbar.

ISBN 978-3-8325-3951-1
ISSN 2364-4877

Logos Verlag Berlin GmbH
Comeniushof, Gubener Str. 47,
10243 Berlin
Tel.: +49 (0)30 42 85 10 90
Fax: +49 (0)30 42 85 10 92
INTERNET: http://www.logos-verlag.de

Contents

Acknowledgements

A great many number of people have been behind the successful completion of this work and I would like to express my appreciation for them.

Firstly, my sincere gratitude to Prof. Dr. Erwin Flaschel for inviting me to work in his department, for the unflinching support and motivation while guiding me through this project and for all the invaluable discussions over the past four and a half years that have played an important role in moulding my competence in the field of bioprocess as well as scientific research in general. I also thank him for the diligent scrutiny of this thesis and significant feedback.

Prof. Dr. Karl Friehs has been an amazing supervisor who kept reminding me about the importance of perseverance and composure when trying to modify genes. I thank him for the support and for his faith in me.

Sincere thanks to Prof. Dr. Thomas Noll for kindly accepting the spontaneous request to take up the role of evaluator for this thesis.

Dr. Gerhard Miksch, who passed away in 2011, was an excellent colleague and friend. He helped immensely through my initial stages in the molecular biology laboratory and would always be deeply missed.

My thanks to Daniela for her kind support all the way. Many thanks to Usama for data regarding the background work with plasmid p582. My thanks to Joe for help with the HPLC, for critical reading of the work and for his support in many other ways. My heartfelt thanks to Thomas and Ebson for making life seem not so hard in the lab. Sincere thanks to all my friends at D5-Fermentation Engineering: to Waleri for all the wonderful discussions, to Maurice, Philipp, Daniel and Lars for all the help and support during those initial days. Thanks to Thomas and Jan for many useful discussions and for being nice office mates. Thanks to Yingfei for help with IMAC, to Jakob for discussions on fluorescence measurement, to Kirsten, and Katrin for storage and maintenance of the strains, to Martin, Dominik, Galina, Thorsten and Heinrich for their help in the lab and to all others for generally being amazing people to work with. A special word of appreciation to Sabine, Franziska, Bastian, Semsinur, Asmae and Vanessa for their work in various projects with my guidance. At the CeBiTec, I would like to thank Prisca for the sequencing, and Christian and Inga for excellent discussions on RT-qPCR.

I would like to thank the German Academic Exchange Service for the PhD scholarship and also for sponsoring the language course which played a crucial part in helping me to learn about and integrate into german society. I thank the Chair of Fermentation Engineering for financial support for the research as well as for sponsoring participation in various conferences where I have had the opportunity to meet and interact with the leading people from the field of bioprocess engineering.

My friends Prasanth (Macquarie University, Sydney) and Premnath (University Hospital of Lausanne) have been amazing people to discuss ideas and share experiences with. Harish (DKFZ, Heidelberg) has been a very supportive friend and a great host.

My parents have been a constant source of support and encouragement throughout and deserve my sincere gratitude. Many thanks to my brother Krishna and my sister-in-law Vidya for their care and encouragement.

Research work usually comes with sacrifices, and often people who are the closest bear the brunt of that experience. Sincere and heartfelt thanks to my partner Sarah for her patience and support.

List of symbols

c_{etPr}	total protein concentration in extracellular medium, g L^{-1}
c_{nPl}	eluted plasmid concentration normalized to optical density of sample, ng μL^{-1}
C_r	concentration of copies of target sequence, copies reaction^{-1}
C_T	threshold fluorescence cycle, -
D	space velocity during chemostat operation, h^{-1}
D_{rel}	relative space velocity, -
$D_{rel,1}$	ratio of D_{rel} to s_{0S} defined as the relative space velocity parameter, -
Da	Damköhler number, -
D_N	space velocity for a cascade of stirred tanks, h^{-1}
DO	dissolved oxygen, % saturation
E	recombinant β-glucanase activity in extracellular medium, U mL^{-1}
E_x	efficiency of PCR amplification of target x, -
f	relative residual substrate concentration at steady state, -
F	volumetric feed flow rate, L h^{-1}
F_n	fluorescence normalized to optical density, -
g	gravitational acceleration, m s^{-2}
k'	ratio of μ_{max} to K_S, L g^{-1} h^{-1}
k_d	cell death rate constant, h^{-1}
K_S	Monod constant, g L^{-1}
$L_{V,E}$	volumetric recombinant β-glucanase productivity, kU L^{-1} h^{-1}
m	slope of C_T versus $\log_{10}$ of template concentration plot, -
m_S	maintenance coefficient, g g^{-1} h^{-1}
μ	specific growth rate, h^{-1}
μ_{max}	maximum specific growth rate, h^{-1}
N_n	number of amplicons after n PCR cycles, -
N_p	copy number of plasmid in a cell, -
OD_{600}	optical density at 600 nm, -
Φ	relative biomass concentration at steady state, -
P_{rel}	relative plasmid abundance with respect to genome, -
q_S	specific substrate consumption rate, h^{-1}
r_S	rate of substrate utilization, g L^{-1} h^{-1}
r_X	rate of biomass increase, g L^{-1} h^{-1}
S_0	concentration of limiting substrate in feeding solution, g L^{-1}
S	concentration of limiting substrate in reactor, g L^{-1}
s_{0S}	saturation parameter, -
S_{AS}	residual ammonium sulphate concentration in medium, g L^{-1}
S_{Gly}	residual glycerol concentration in medium, g L^{-1}
$S_{P/X}$	selectivity of recombinant enzyme over biomass, g g^{-1}
t	batch/ fed-batch/ chemostat operating time, h
t_d	doubling time, h
T_m	melting temperature, °C
θ	probability of appearance of plasmid-free daughter cell, -
V	volume of culture medium, L
X	dry biomass concentration, g L^{-1}
$Y'_{X/S}$	true biomass yield coefficient over substrate, -
$Y_{X/S}$	apparent or observed biomass yield coefficient over substrate, -

List of subscripts

0	initial condition
600	wavelength in nm
AS	ammonium sulphate
c	critical
etPr	extracellular total protein
E	enzyme
Gly	glycerol
i	any vessel in a cascade
I,0	first kind, zero order
max	maximum level
n	number of PCR cycles
opt	optimal
p	plasmid
P/X	product with respect to biomass
rel	relative
rel,1	relative, first order
S	substrate
T	threshold
V	volumetric
x	target sequence
X	biomass
X/S	biomass with respect to substrate

List of abbreviations

AP	Alkaline Phosphatase
BRP	Bacteriocin release protein
CCC	covalently closed circular
cDNA	complementary DNA
CFU	colony forming unit
CSTR	Continuous Stirred Tank Reactor
DEPC	Diethylpyrocarbonate
DHAP	Dihydroxyacetone phosphate
DNA	Deoxyribonucleic acid
dNTP	deoxynucleoside triphosphate
DP	Degree of Polymerization
EC	Enzyme Commission
EDTA	Ethylenediaminetetraacetic acid
G3P	Glyceraldehyde 3-phosphate
GAPDH	Glyceraldehyde 3-phosphate dehydrogenase
GFP	Green Fluorescent Protein
IB	Inclusion body
IFN	Interferon
IMAC	Immobilized metal ion affinity chromatography
IPTG	Isopropyl-β-D-thiogalactopyranoside
LB	Lyogeny Broth
MCS	multiple cloning site
mRNA	messenger RNA
NTA	Nitrilotriacetic acid
NTC	no template control
OM	Outer Membrane
OMPLA	Outer Membrane Phospholipase A
ORT	Operator Repressor Titration
PCR	Polymerase chain reaction
ppGpp	Guanosine tetraphosphate
qPCR	quantitative Polymerase Chain Reaction
RNA	Ribonucleic acid
rRNA	ribosomal RNA
RT	Reverse transcriptase
RT-qPCR	Reverse Transcription-quantitative Polymerase Chain Reaction
SDS-PAGE	Sodium dodecyl sulphate-polyacrylamide gel electrophoresis
SG	SYBR Green I dye
SGA	Supplemented Glycerol Ammonium sulphate medium
TAE	Tris-acetate EDTA
TB	Terrific Broth
TCA	Trichloroacetic acid
TEMED	N-N-N´-N´-tetramethylethylenediamine
tRNA	transfer RNA
UV	ultraviolet
bla	beta-lactamase
bp	basepair
ds	double stranded

fic	filamentation induced by cyclic adenosine monophosphate
i.d.	internal diameter
mA	milliampere
mol	mole
nm	nanometer
R	resistant
rpm	revolutions per minute
vvm	volume of air per volume of medium per minute
w.r.t	with respect to
x	times

1 Introduction

The tremendous amount of research carried out in the second part of the last century has resulted in the taming of a mostly benign, occasionally pathogenic gut bacterium *Escherichia coli* into a safe, robust, well-understood and optimized strain for recombinant production towards various biotechnological objectives. These range from production of industrial enzymes and bioethanol to eukaryotic therapeutic proteins and DNA vaccines. A classic example is the production of recombinant human insulin in *E. coli* in the late 1970s. Today, although a wealth of knowledge on the genetic engineering strategies and bioprocess methods for efficient recombinant production with *E. coli* exists, it has hardly stopped the flow of new ideas for improvement. One such key area of interest has been to find ways of getting around a basic deficiency in *E. coli* – the inability to efficiently excrete recombinant proteins into the culture medium. Localization of the foreign protein within the cells could render them a target for degradation by cellular proteases, and overproduction may result in the formation of inactive aggregates, whereas excretion into the medium would greatly benefit downstream purification. In this regard, the controlled co-expression of the bacteriocin release protein (BRP) to release recombinant proteins that have been previously targeted to the periplasmic space was a strategy developed in this research group and is now a fairly established method. However, this strategy has until now only been studied under batch and fed-batch conditions, and this doctoral thesis primarily deals with the establishment and then analysis of this concept extended to the continuous mode of operation. Overproduction of recombinant protein is often favoured at distinct specific growth rates. Continuous culture - due to its flexibility for control and reduced down-time - could in many cases offer higher productivity than batch or fed-batch processes (Bull, 2010). Particularly, where it could be coupled with extracellular protein secretion, continuous cultures deserve attention (Shokri *et al.*, 2002).

Once established, the stable operation of continuous cultures is of utmost importance in order to be able to deliver a stable volumetric productivity. Structural and segregational stability of the recombinant plasmid come into the picture at this point. Whereas the former can be controlled to an extent by the choice of the host strain, the latter is usually maintained by employing a selection pressure for maintenance of the plasmid. Antibiotics are a common method of achieving this selection pressure in genetic engineering and small-scale cultivations but their use is both uneconomical and unsustainable in large-scale bioprocesses and in long-term continuous cultures. Therefore, an alternative antibiotic-free method of maintaining the recombinant plasmid in the cells had to be developed which forms the second major part of this thesis. In this regard, two different metabolic pathways have been targeted to create an auxotrophic complementation system that makes the cell dependent on the recombinant plasmid for survival under the applied conditions.

In the final part, the activity of a growth-phase regulated promoter under continuous culture conditions is studied in detail at the transcriptional level by employing reverse transcription-quantitative real-time PCR and green fluorescent protein transcriptional fusion.

2 Theory

2.1 Background information

2.1.1 *Escherichia coli* – the 'workhorse'

Already in 1977, in a book chapter, Kennedy and Dixon regarded *Escherichia coli* as the "longstanding workhorse of molecular genetics" (Kennedy & Dixon, 1977). Since then, this term has been used to denote *E. coli* in numerous publications, and it is not without reason. In 2009, 30% of recombinant biopharmaceuticals were produced in *E. coli* (Ferrer-Miralles *et al.*, 2009) and in 2013, 73% of recombinant genes were expressed in *E. coli*, which, although it represents a slight decline over the last 8 years, still underscores the fact that *E. coli* as a host remains an important microorganism for both industry and academia (Bill, 2014). A number of reviews in the past have documented the principles and strategies for optimally exploiting this microbe for the production of recombinant products (Baneyx, 1999; Sørensen & Mortensen, 2005; Jana & Deb, 2005; Terpe, 2006; Huang *et al.*, 2012). Studies like the use of N^{pro} fusion technology only prove that there are still immense opportunities to develop and optimize the use of this microorganism (Achmüller *et al.*, 2007). Although they are excellent for expression of native proteins, the use of Gram-positive hosts such as *Bacillus* sp. for the production and excretion of heterologous proteins is hampered by the large number of proteases that are also excreted by these strains (Simonen & Palva, 1993), wherein *E. coli* presents a significant alternative. The attractiveness of this expression system for recombinant proteins can therefore be greatly enhanced by adding an efficient secretion route for the product (Ni & Chen, 2009).

2.1.2 Membrane structure

Escherichia coli is a rod-shaped bacterium with a cellular envelope typical of a Gram-negative bacterium as shown in Fig. 2.1. The cytoplasmic contents are confined by a phospholipid bilayer called the cell or inner membrane (IM) which is enveloped by a characteristic thin layer of peptidoglycan. This second layer is in covalent linkage to lipoproteins of the outermost layer called the outer membrane (OM) which is again a distinguishing feature of Gram-negative bacteria. The OM is an asymmetrical membrane composed of phospholipids (PL) in the inner leaflet and lipopolysaccharides (LPS) in the outer leaflet (Ruiz *et al.*, 2009).

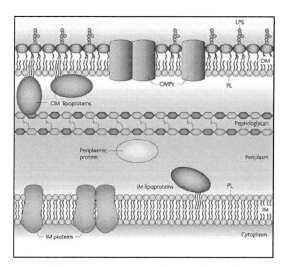

Fig. 2.1: Cellular envelope of a Gram-negative bacterium. Image adapted from Ruiz *et al.*, 2009 (Copyright Macmillan Publishers Limited).

2.2 Extracellular transport

In this work, the term secretion would be used to denote the transport through the cell or inner membrane resulting in accumulation of the target protein within the space between the inner and outer membranes called the periplasmic space. Further permeation through the outer membrane and release into the extracellular medium would be denoted by the term excretion. The periplasmic space holds great physiological significance since it is in this oxidizing environment in the presence of important chaperones and foldases that disulfide bond formation and isomerization are promoted, and many secreted proteins are folded to their correct and final conformations (De Marco, 2009). Due to these and other reasons like fewer contaminating proteins and therefore easier purification and avoiding proteolytic degradation by intracellular proteases, it is very advantageous to secrete a recombinant protein through the inner membrane (Yoon, 2010). Usually, recombinant proteins can be directed to the secretory route by addition of a short signal sequence which is cleaved away during translocation into the periplasm. Some commonly used signals among others are those from the proteins PelB, OmpA, OmpC, PhoA and MalE (Choi & Lee, 2004).

In *E. coli*, the type I and type II are the most commonly used excretory pathways for recombinant proteins. The type I excretion mechanism, best characterized in the α-hemolysin (HlyA) transporter, involves the 2 inner membrane proteins HlyB and HlyD which form a complex with the recombinant protein, and together with the outer membrane protein TolC, form a direct channel for export from cytoplasm to the extracellular medium (Mergulhão *et al.*, 2005). A recombinant lipase B in active form has been efficiently exported to the extracellular medium through the HlyA fusion route and also by using the modified flagellar (type III) protein secretion route (Narayanan *et al.*, 2010). One main problem with the HlyA fusion route is that the target protein has to be cleaved away from its fusion partner in a separate downstream step.

In the type II pathway, (the general secretory pathway) proteins may be excreted in a two-step process, involving passage through the cell membrane to reach the periplasm, followed by permeation of the OM. The first part of the transfer may be possible through either of three different routes:

The Sec pathway is the most common route wherein the protein is bound post-translationally by SecB and delivered to the SecYEG complex through SecA and then translocated through the cell membrane.

The major difference in the second route is that the signal recognition particle (SRP) in complex with the ribosome, recognizes its substrate protein resulting in the co-translational binding to Ffh and subsequent delivery to the translocation site on the membrane.

The third route is the twin-arginine translocation (Tat) system involving proteins TatA, TatB, TatC and TatE, secretes proteins already folded in the cytoplasm, and has also been used for secreting various recombinant proteins (Mergulhão et al., 2005).

Thus, E. coli generally stops at this level and does not excrete through the OM. After secretion to the periplasm, the main terminal branch of the general secretory pathway is responsible for extracellular transport. Although, genes coding for this system are present in E. coli, they are silenced by the H-NS protein under normal laboratory conditions (Francetic et al., 2000).

The limited extracellular secretion of native proteins by E. coli (Hannig & Makrides, 1998) represents a major difference to Gram-positive hosts such as Bacillus subtilis which frequently secrete into the extracellular medium (Harwood & Cranenburgh, 2008). This could also be a major advantage in the case of E. coli since if extracellular transport could be achieved, it would result in the recombinant protein forming the major component of the protein mixture in this fraction (Mergulhão et al., 2005). Leakage of periplasmic proteins into the culture medium at size-dependent rates and the potential to exploit this phenomenon to recombinant proteins has been reported (Rinas & Hoffmann, 2004). Another recent study described mutations in the N-terminal region of a subtilase propeptide that improved secretion into the periplasm through the SecB pathway, but the subsequent leakage into the extracellular medium remained unexplained (Fang et al., 2010).

A variety of passive transport routes for periplasmic proteins to traverse the outer membrane including mechanical, chemical or enzymatic methods have been reviewed by Shokri et al. (2003). Leaky mutants, strains with defective outer membrane structures due to deficient lipid A biosynthesis have been studied in the past for their ability to release periplasmic enzymes into the extracellular medium (Nurminen, et al., 1997). Leaky strains carrying mutations in genes including omp, tol, env, lky, excD and excC have been successfully applied in the industry for the extracellular production of correctly folded and assembled functional full-length IgG and IgM antibodies (European patent EP1903115 B1, 2011). Recently, there has been extensive interest in developing strains like lpp deletion mutants which lack the murein lipoprotein in the OM and, therefore, exhibit an enhanced permeability to release recombinant proteins from the periplasm into the extracellular space with only minor loss in cell growth rate (Shin & Chen, 2008).

The other major group of strategies for the release of recombinant proteins from the periplasm into the extracellular medium is the internal lysis method. Plasmids like pColE1 and pCloDF13 coding for bacteriocins also include genes coding for bacteriocin release proteins (BRP) – small lipoproteins necessary for excretion of the bacteriocin – in their gene clusters. Full-induced expression of BRPs could cause decrease in culture turbidity referred to as 'quasi-lysis'. However, low-level expression of BRP causes a semi-selective increase in outer membrane permeability which results in the release of periplasmic proteins into the extracellular medium (Van der Wal *et al.*, 1995a, Miksch *et al.*, 1997a).

Precursor polypeptides of BRPs range from 42 to 57 amino acid residues in length, and contain the Leu-X-Y-Cys consensus sequence at the carboxy end of the signal peptide cleavage site where X and Y are Val and Gly respectively in the case of ColE1 BRP. The 4.5 kDa precursor is transported through the IM over the Sec pathway, undergoes a rapid lipid modification, followed by cleavage of the 17 residues long signal peptide by signal peptidase II, prior to reaching the OM. The Cys in the consensus sequence is the site for lipid modification which is necessary for both correct targeting and functioning of mature BRP (Cascales *et al.*, 2007). The mature colicin E1 release protein relevant in this study has a length of 28 amino acid residues and as an exceptional case, gets processed within seconds (Van der Wal *et al.*, 1995a).

For bacteriocins, although the exact mechanism of release through both membranes is not known, it is thought that due to the concerted action of lipid-modified BRP, its stable signal peptide residing in the Sec complex and the action of phospholipase A, a trans-envelope pore is realized that results in release of the bacteriocin to the extracellular medium (Van der Wal *et al.*, 1995a, Cascales *et al.*, 2007). However, for recombinant protein expression, the release of periplasmic proteins due to the subsequent activation of outer membrane phospholipase A (OMPLA) in regions on the OM other than the putative trans-envelope pores is the relevant phenomenon. Therefore, it is advantageous to have an unstable SP that would avoid the unnecessary formation of these pores so that only OM perturbation would be caused (Van der Wal *et al.*, 1995a; Luirink *et al.*, 1991). Therefore the unusually unstable signal peptide of ColE1 BRP which is rapidly degraded, is advantageous. This also has the advantage that the Sec transporters would be freed from clogging by stably bound SPs (Sommer *et al.*, 2010) thus making them available for transport of the target recombinant protein.

It has been proposed that the perturbation of the OM caused by BRP results in changes in its asymmetrical nature and an increased permeability. This allows the inactive OMPLA monomers to come closer and dimerize, resulting in their activation (Dekker *et al.*, 1999). Additionally, OMPLA is calcium-dependent for dimerization and catalytic activity (Snijder & Dijkstra, 2000). The sequence of events leading to OMPLA activation and OM structure disruption is illustrated in Fig. 2.2. Normally, OMPLA is an integral OM protein present as inactive monomers with a bound calcium ion (A).

Mature BRP creates a perturbation in the form of incorporation of phospholipids in the outer leaflet which increases fluidity of the OM and causes lateral diffusion of its integral proteins (B). Additionally, substrates are presented to the OMPLA monomers, dimerization takes place and the calcium ion is thought to move to the active site creating an 'active complex' made of OMPLA dimers, substrates and calcium ion as cofactor (C). The active complex hydrolyzes the phospholipids in the OM and generates lysophosphatidylethanolamine as the major lysophospholipid and free fatty acids which can further permeate through the membrane (D). These hydrolysis products destabilize the membrane and cause loss of bilayer structure. Non-bilayer structures can increase the OMPLA activity up to 70-fold leading to even faster hydrolysis of membrane phospholipids, thus amplifying the membrane destabilizing effect. The resulting loss in OM structure allows periplasmic proteins to be released into the extracellular medium (E) (Snijder & Dijkstra, 2000).

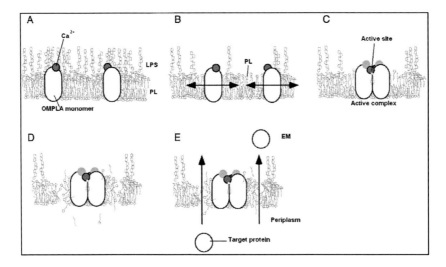

Fig. 2.2: Mechanism of OM permeabilization and periplasmic protein release following OMPLA activation. LPS: lipopolysaccharide, PL: phospholipid, EM: extracellular medium. Refer to text for details. Image adapted from Snijder & Dijkstra, 2000. Copyright © 2000, Elsevier.

The successful secretion of a periplasmic xylanase of molecular mass 135 kDa from *Aeromonas* sp. represents the scale of size of target proteins that are compatible with the BRP co-expression strategy for extracellular expression (Kato *et al.*, 1986). Other examples of recombinant proteins excreted using BRP co-expression include phytase (Miksch *et al.*, 2002), penicillinase (Aono, 1989) and α-amylase (Yu & San, 1992).

The *kil* gene of pColE1 coding for BRP was co-expressed from a growth-phase-dependent promoter and tested for the release of heterologous proteins from the periplasm into the extracellular medium (Miksch *et al.*, 1997a) thus presenting a strategy that forms the major foundation of this work. The use of the weak stationary-phase promoter of the *fic* gene for control of *kil* allowed regulation of strength, time and automatic BRP activation for extracellular expression of a recombinant hybrid β-glucanase while avoiding quasi-lysis and lethality. In addition, an observed overproduction of the target enzyme in cells expressing BRP compared to *kil* control cells, led to the assumption that the emptying of the periplasmic space relieved stress and promoted more synthesis (Miksch *et al.*, 1997a). Optimal promoter distance from the *kil* gene, growth medium, and bioprocess conditions for the production of this target enzyme were further established using *E. coli* JM109 as host strain (Miksch *et al.*, 1997b). Thus, it was found that a shorter distance between the promoter and the *kil* gene was beneficial for better product excretion. Other factors such as the addition of 1-2% NaCl or increased oxygen supply improved enzyme activity levels as well.

2.3 Model protein – industrial enzyme β-glucanase

The cell walls in the bran and starchy endosperm of cereal grains of the family Poaceae such as barley or oats contain linear unbranched and unsubstituted (1,3;1,4)-linked β-glucans which play a structural role by forming a gel-like matrix between the cellulose microfibrils. In barley, the glucosyl residues may reach a DP of over 1000. Outside of angiosperms, (1,3;1,4)-β-glucans have also been found in a few algae, bryophytes and monilophytes but studied in most detail in the lichen *Cetraria islandica* (Icelandic moss) which is a composite organism from a symbiotic relationship between a fungus and a photobiont (Harris & Fincher, 2009). The polymer structure for barley β-glucan is shown in Fig. 2.3. A pattern of individual (1,3)-linkages separated by 2 or 3 (1,4)-linked glucosyl residues are frequent although contiguous units of up to a DP of 14 are also present (Stone, 2009).

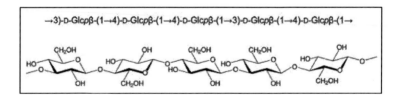

Fig. 2.3: Structure of (1,3;1,4)-β-glucan polymer. The designation 1,3 or 1,4 refers to the linkage between the carbon atoms of adjacent monosaccharide units with the designation β referring to the fact that the oxygen bridge forming the link is present above the ring of the β-glucopyranosyl monomer. Image adapted from Stone, 2009. © 2009, Elsevier Inc.

As a model protein of industrial significance, a hybrid β-glucanase from *Bacillus* sp. (Borriss *et al.*, 1989) has been expressed in this work. During beer production, endogenous hydrolases of barley are invariably heat-inactivated during kilning and mashing (Planas, 2000). Insufficiently depolymerized β-glucans prove a challenge to the brewing industry by increasing solution viscosity, reducing filtration rate and causing beer haze in the final product thus making it necessary to add thermostable bacterial β-glucanases. The presence of β-glucans in feed formulations slows down digestion and reduces growth rate of animals (Harris & Fincher, 2009). Correspondingly β-glucanases have been shown to be useful in a variety of applications in white biotechnology like brewing (Celestino *et al.*, 2006), animal feed additives (Boyce & Walsh 2007; Choct, 2006) and in paper and pulp processing (Gil *et al.*, 2009). For the detergent industry, β-glucans form a major portion of the non-starchy polysaccharides of soils on textiles, requiring the application of β-glucanases in addition to the usual enzymes (International patent WO1999006516 A1, 1999).

The endo-1,3-1,4-β-D-glucan 4-glucanohydrolases (EC 3.2.1.73) are specific for the cleavage of a (1,4)-β-D-glucosyl linkage in the presence of a (1,3)-β-D-glucosyl residue on the non-reducing end. Interestingly, although microbial and plant 1,3-1,4-β-glucanases have identical substrate specificities, they share neither primary nor tertiary protein structures. The microbial enzymes also called Lichenases, are classified under the Family 16 of glycoside hydrolases and show a jellyroll antiparallel β-sheet sandwich structure (Fig. 2.4). The mechanism of hydrolysis is a double-displacement reaction catalyzed by 2 essential residues, one acting as a general acid/base and the other as a nucleophile (Planas, 2000).

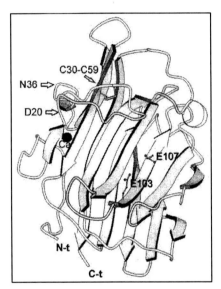

Fig. 2.4: Three-dimensional structure of *Bacillus macerans* 1,3-1,4-β-glucanase showing the side-chains of catalytic residues E103 and E107, a surface loop from residues 20-36 partially covering the substrate binding cleft, the disulphide bond between cysteines 30 and 59 connecting this loop to the core and the calcium ion bound at a region opposite to the active site. Image adapted from Planas, 2000. © 2000 Elsevier Science B.V.

High induction by temperature shift was used to drive the expression of a *Bacillus amyloliquefaciens* β-glucanase in *E. coli* using the P_R promoter and C_{I857} temperature-sensitive repressor by Riethdorf *et al.* (1990). This even lead to excretion of the periplasmic target enzyme into the growth medium which was postulated to have been caused by its overexpression. Recently, various *Bacillus* enzymes were expressed and secreted in *E. coli* using native or *E. coli* OmpA signal peptides. Though the mechanism was not known, time after induction, type of signal peptide and the size of the recombinant protein were found to be factors influencing extracellular release of the enzymes (Yamabhai *et al.*, 2008). The bioethanol industry is set to face increasing demands to shift from starch-based resources towards non-starch polysaccharides in order to address the food vs. fuel issue. In this context, non-starch polysaccharides other than cellulose and hemicellulose need to be looked into. In this regard, the use of bacterial endo-1,3-1,4-β-glucanases along with β-glucosidases for the breakdown and saccharification of barley β-glucan and subsequent yeast fermentation to produce ethanol represents an interesting application for β-glucanases (Divate *et al.*, 2013).

2.4 Plasmid selection and alternative strategies

Plasmid-based microbial processes depend heavily on plasmid segregational and structural stability as well as plasmid copy number maintenance. Segregational instability of plasmids can be mathematically related to the copy number in a cell. Thus, the probability (θ) that a plasmid-free daughter cell arises from a plasmid-bearing cell in the absence of any regulation is given by $2^{(1-N_p)}$ where N_p is the plasmid copy number (Blanch & Clark, 1996).

Plasmid-free cells can lead to a decrease in the overall product recovery and profitability of a recombinant bioprocess (Kroll *et al.*, 2010). Several natural plasmid maintenance systems employed by bacteria have been described. These include the *cer* locus of pColE1 active in multimer resolution and thereby reducing the probability of plasmid-less progeny and the *par* locus of pSC101 that actively distributes low-copy number plasmids to the daughter cells. However, these systems are not very effective when dealing with the challenges of an industrial bioprocess using a recombinant plasmid and more importantly for a continuous culture process (Porter *et al.*, 1990).

Apart from the plasmid distribution among daughter cells, other factors such as recombinant gene induction and expression (Hellmuth *et al.*, 1994), growth environment (Gupta *et al.*, 1995), size of the plasmid, copy number, formation of multimers and growth advantage of plasmid-free cells (Friehs, 2004) could become critical issues. Typically, there is a distribution of plasmid copy numbers among the cells in a population. The risk of plasmid loss from the culture in a bioprocess becomes particularly serious when a toxic or highly energy-draining heterologous gene is expressed. This is because, plasmid-free cells arising as a result of copy number distribution, will now have a marked growth rate advantage over plasmid-bearing cells and will eventually take over the culture resulting in loss of productivity (Friehs, 2004).

One of the most common methods of maintaining recombinant plasmids in the host cell is to add antibiotics to the culture medium and make use of a cloned resistance gene on the plasmid to create a selection pressure in favour of plasmid-carrying cells. However, this idea is restricted to small-scale laboratory cultivations aimed at studying the molecular genetics or physiology of the organism. For large-scale processes and especially for extended continuous cultivations, plasmid maintenance cannot be achieved efficiently by using antibiotics/antibiotic-resistance genes due to a variety of reasons such as high cost of antibiotics, the need to remove them in downstream processes, and due to regulatory stipulations for food and pharmaceuticals (Friehs, 2004). Apart from this, the spread of antibiotic-resistance genes in the environment and the emergence of multiresistant pathogens are a cause of concern in recent times (Glenting *et al.*, 2002). There is also the possibility that the selection pressure may be lost over time as a result of degradation of the antibiotic, a problem particularly acute with ampicillin (Friehs, 2004; Vidal *et al.*, 2008). A two-plasmid T7-based expression system with *E. coli* OmpA signal sequence for secretion was used to express the toxic product streptokinase. In continuous culture, space velocity was used as a tool to improve expression, and high antibiotic selection pressure was found necessary to tackle plasmid instability. In addition, the system had to compromise with an unsteady state following induction due to the principle involved for induction and recombinant protein expression (Yazdani & Mukherjee, 2002). Strategies developed to ensure segregational stability of plasmids in recombinant cultures, range from molecular genetic approaches to process-based approaches (Friehs, 2004). As an example of a process-based strategy, on a population level, space velocity as a parameter has been used as a tool to maintain the plasmid-bearing cell fraction by periodic operation of a continuous culture (Stephens *et al.*, 1992).

Among the molecular genetic approaches, certain plasmid selection strategies targeted essential components of basic cell physiology and metabolism whereby the defect on the genome was complemented by the corresponding function presented on the plasmid. Examples include the *infA* gene coding for translation initiation factor IF1 (Hägg *et al.*, 2004), the *valS* gene coding for valyl-tRNA synthetase (Nilsson & Skogman, 1986) and the *ssb* gene coding for single-stranded DNA binding protein in *E. coli* (Porter *et al.*, 1990). Yet another group of strategies used complementation by amber suppressor tRNA as a route to antibiotic-free plasmid maintenance. A plasmid design called pCOR used the expression of a phenylalanine-specific amber suppressor tRNA gene, allowing the *argE* mutant host to grow on minimal medium lacking arginine. Furthermore, stringent dissemination control was achieved through the incorporation of an essential trans-acting factor - the π protein from the host genomic *pir* gene which was required by the conditional origin of replication (ori γ) on the plasmid (Soubrier *et al.*, 1999). Another RNA-based strategy is the use of an amber nonsense mutation introduced into the essential *thyA* gene in the chromosome causing thymidine auxotrophy. This was overcome by recombinant plasmids carrying a histidine suppressor t-RNA, which allowed antibiotic-free plasmid selection and also recombinant luciferase reporter expression in eukaryotic tissues and in tumour cells (Marie *et al.*, 2010).

However, the major molecular genetic approaches could be broadly classified into 3 groups – toxin/anti-toxin systems, operator-repressor titration (ORT) systems and metabolism-based systems (Kroll *et al.*, 2010) which are shown schematically in Fig. 2.5.

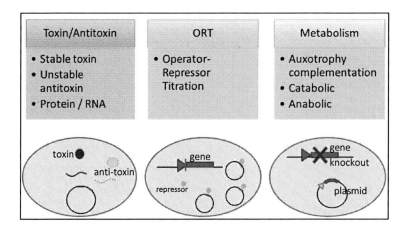

Fig. 2.5: Categorization of molecular genetic strategies for alternative plasmid selection.

a) <u>Toxin/anti-toxin systems</u>: Under the first category, probably the most important strategy is the hok/sok system of plasmid R1, wherein a stable mRNA *hok* (host killing) that encodes a toxic protein, is indirectly prevented from being translated by the binding of a complementary unstable antisense RNA called *sok* (suppressor of killing) which binds to the *mok* (mediation of killing) reading frame leading to RNase III-dependent degradation of the RNA-duplex. Together, due to the differential decay rates, the two RNAs enforce a mechanism of post-segregational killing on plasmid-free cells (Thisted *et al.*, 1994). The effectiveness of this system for plasmid maintenance along with recombinant β-galactosidase expression has been tested in shake-flask cultures. However, diminishing of the selection pressure with increasing medium complexity or level of induction of the cloned gene was also observed. The competition for and shortage of ribosomes to instantaneously translate the killer mRNA and maintain toxin levels to kill plasmid-free segregants was suggested as a possible scenario for the instability observed during high recombinant gene expression. (Wu & Wood, 1994). The *ccdB/ccdA* system consists of a stable toxic protein CcdB that binds to DNA gyrase and prevents DNA replication, and its natural inhibitor, the unstable CcdA antidote. Stable expression of recombinant *Pseudomonas* exotoxin was demonstrated by the use of a strain with a genomic copy of *ccdB* and a plasmid-borne copy of *ccdA* together with multimer resolution made possible by a cloned *cer* locus (Peubez *et al.*, 2010). A critical issue affecting the desirability of toxin/anti-toxin systems is that they have also been implicated to be frequent among plasmids encoding extended spectrum beta-lactamases in *E. coli* strains (Mnif *et al.*, 2010).

Many studies on plasmid DNA production for gene therapy have particularly been interested in alternative plasmid selection mechanisms, due to the need to avoid all kinds of antibiotic resistance genes or their products in therapeutic DNA or DNA vaccines in conformation with regulatory stipulations. A method based on the expression of a plasmid-coded antisense RNA to repress a constitutively expressed genome-based counter-selectable marker *sacB* during growth on sucrose was reported recently to be able to bring about antibiotic-free selection and highly productive fermentation while not being restricted to ColE1 vectors (Luke *et al.*, 2009). An interesting strategy to stabilize ColE1 plasmids without the need for any additional modification was shown to be possible by modifying the host *E. coli* strain. The RNA I from the origin of replication of the plasmid binds to and inhibits the translation of the mRNA from *tetR* gene that has been modified with RNA II sequences. In the absence of the plasmid, the TetR protein binds to the tet operator and represses the expression of the adjacent essential gene *murA* (cell wall biosynthesis). Notably, the tac promoter driving the *tetR* gene was leaky and certain unexpected observations with regard to IPTG induction were also reported (Mairhofer *et al.*, 2008).

b) Operator-repressor titration system: The ORT strategy (Fig. 2.6) is novel in that it represents a non-expressed selection marker. It is based on negative regulation of a chromosomal gene *dapD* (essential for diaminopimelate/lysine biosynthesis), by the lac operator sequences O_1 and O_3 allowing the binding of the constitutively expressed lacI repressor. In order to allow expression of *dapD*, the cell has to titrate the repressor molecules against a similar operator sequence that may be present in multiple copies on the target plasmid to be maintained which could even be of low copy number type (Cranenburgh *et al.*, 2004). The advantage of using the *lac* system is that the untransformed host strain can still be manipulated by growing in the presence of an inducer such as IPTG. The effectiveness of this system was later confirmed in a process for high cell density cultivation and recombinant fuculose-1-phosphate aldolase production (Durany *et al.*, 2005).

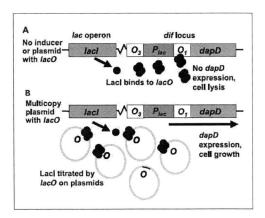

Fig. 2.6: Representation of the operator-repressor titration strategy for plasmid selection and maintenance. Image taken from Cranenburgh *et al.*, 2004. © 2004 S. Karger AG, Basel.

The fabI-triclosan system, which is an essential gene-gene product inhibitor combination, is a completely different plasmid selection concept and incorporates features of both toxin/anti-toxin as well as repressor titration categories. However, aspects like the requirement of the selection agent due to an addictive effect, the need to induce and over-express the essential gene marker, and plasmid instability in the absence of selection were also noted. Although triclosan is an approved biocide, some of the risks associated with its use tend to go back towards those arising from the use of antibiotic resistance genes (Goh & Good, 2008).

c) Metabolism-based systems: Strategies using this approach could target either catabolic or anabolic pathways. The basic principle involves creating a mutant strain that contains a loss of essential function on its genomic locus which requires the strain to be cultivated under specific conditions/medium. The auxotrophy in the genome is complemented by a copy of the deleted gene cloned onto the recombinant plasmid under the control of an appropriate promoter. The selection pressure is now provided by the complementing gene which renders the plasmid absolutely essential for the cell to be able to survive under the applied conditions. Catabolism-based auxotrophic systems target an essential step in the pathway for utilization of a substrate. One critical aspect with these systems though, is the need to be restricted to a defined medium containing one specific source of carbon and energy in order to maintain stability. Furthermore, due to the versatility of the *E. coli* metabolic network, catabolic systems have more chances of bypassing the auxotrophy through alternate pathways (Kroll *et al.*, 2009).

On the other hand, a large number of anabolism-based systems have been reported in the literature. The gene coding for quinolinic acid phosphoribosyltransferase (QAPRTase) catalyzing an important step in the *de novo* nicotinamide adenine dinucleotide (NAD) biosynthesis pathway has been deleted in the genome and supplied on a plasmid by replacing the antibiotic resistance gene to result in an efficient plasmid selection and expression system (Dong *et al.*, 2010). An extremely sophisticated anabolism-based auxotrophic pathway was reported for the stabilization of a plasmid expressing cyanophycin synthetase for production of cyanophycin. A knock-out of the *ispH* gene (involved in synthesis of isopentenyl pyrophosphate which is required for cell wall biosynthesis among others) from the *E. coli* HMS174 (DE3) genome was followed by the introduction of a plasmid carrying a completely artificial mevalonate-dependent pathway based on genes from *Lactococcus* sp. and *Staphylococcus* sp.. Although it led to a stable expression system, the resulting final plasmids which were up to 14 kb in size, probably caused extreme stress on the knock-out hosts, in addition to the need to synthesize multiple enzymes in the alternate pathway, resulting in lower growth rates and lower specific recombinant enzyme activity in comparison to control strains (Kroll *et al.*, 2009). Large plasmids also carry a higher risk of structural instability. The group also expressed the same product using a *dapE* deletion host that was made auxotrophic for lysine due to inability to synthesize L,L-2,6-diaminopimelate (LL-DAP). The target plasmid carrying the *dapL* gene from *Synechocystis* sp. resulted in a successful complementation through an alternative pathway and the system was tested under fed-batch conditions in media lacking lysine. Probably due to the diffusion of auxotrophy-related metabolites, the auxotrophic selection pressure, although resulting in stability, could not be sustained over a long period or in high-cell density cultivations (Kroll *et al.*, 2011). Interestingly, even if lysine would be provided externally, the selection pressure should still exist in a *dapE* deletion host since the molecule meso-2,6-diaminopimelate (m-DAP) cannot be synthesized and it would be required for cell wall synthesis. This aspect has been discussed in an earlier work also involving the lysine biosynthesis pathway. Here the *dapD* gene was targeted and a successful auxotrophic system was setup and tested for the stable production of IFN-γ (Degryse, 1991). Therefore, cross-feeding risks for the lysine strategy probably involve the uptake of LL-DAP and/or m-DAP.

There are many other anabolism-based auxotrophic systems dealing with amino acid biosynthetic pathways. One example is the deletion of the *glyA* gene involved in the biosynthesis of glycine through threonine or serine. A complementation plasmid carrying the homologous *glyA* gene under control of a weak constitutive promoter P3 could enable the growth of the auxotrophic strain in minimal medium lacking glycine and was used for the overproduction of rhamnulose-1-phosphate aldolase in antibiotic-free fermentation (Vidal *et al.*, 2008). However, this system has been tested only under batch and fed-batch conditions and the authors point out the need to study such auxotrophic complementation systems under long-term continuous culture conditions. This is especially interesting because there is an escape pathway through threonine, catalyzed by the *ltaE* gene product which could be favoured in mutants that may arise over prolonged cultivation. The production of a partially humanized antibody fragment (Fab) in a proline auxotrophic strain *E. coli* JM83 was possible by including a complementing *proBA* operon to the plasmid (Fiedler & Skerra, 2001). Again, the fact that the complementation operon was controlled by its native promoter and terminator sequences made it possible for recombination with the host genome and reversal of the Δ*proBA* genotype leading to loss of selection pressure. From the leucine biosynthetic pathway, the *leu2* gene is a long-known selectable marker for 2 μm-plasmids maintained in leucine auxotrophic strains of *Saccharomyces cerevisiae* (Sikorski & Hieter, 1989). In bacteria, the *leuD* gene has found application in the development of an auxotrophic system for the stable expression of recombinant antigens from the live vector *Mycobacterium bovis*. A Bacillus Calmette-Guérin (BCG) strain was modified to be auxotrophic for leucine by the knockout of genomic *leuD* gene through homologous recombination. Expression of *leuD* on a complementing plasmid proved it to be an efficient selectable marker in media lacking leucine. However, an extreme instability of the system *in vitro* was seen when the medium contained 100 mg L^{-1} leucine (non-selective) (Borsuk *et al.*, 2007).

Although most parallel strategies that aim at stable genomic integration of heterologous genes result in a gene dosage of a single copy and do not measure up favourably against multi-copy plasmid systems, recently *recA*-mediated multiple target gene copy integration into the genome was shown possible in *E. coli* by chemically induced chromosomal evolution and subsequent stabilization through *recA* deletion. This was successful in antibiotic-free stable maintenance of heterologous gene for producing PHBs and lycopene as model products (Tyo *et al.*, 2009).

2.5 The leucine biosynthetic route

The leucine biosynthetic operon depicted in Fig. 2.7 was one of the targets used in this work for development of an auxotrophic complementation strategy for the stabilization of plasmid p582 without antibiotics.

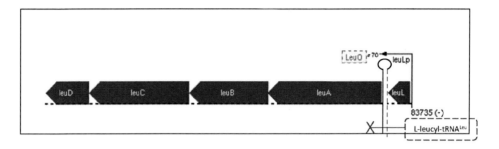

Fig. 2.7: Organization of the *leuABCD* operon. *leuL* is the leader peptide involved in transcriptional control. Image adapted from the online source EcoCyc (www.ecocyc.org; Keseler *et al.*, 2011) © 2014 SRI International.

The *leuABCD* operon consists of a promoter *leuL*$_p$ followed by a transcription control region working on the principle of ribosome-mediated attenuation. This mechanism is based on the simultaneous transcription and translation of a short leader mRNA from *leuL* containing 4 adjacent leucine codons within its sequence. When the cell is deficient in L-leucyl-tRNA^{Leu} the ribosome stalls at this region allowing the nascent RNA to assume a preemptor conformation that prevents formation of a termination loop and hence transcription reads through into the operon. When the cell contains sufficient leucine-charged tRNA, the ribosome completes *leuL* translation and pauses at the stop codon which triggers formation of the termination loop on the mRNA strand and the operon genes are not expressed (Wessler & Calvo, 1981). The transcription control region is followed by the structural genes *leuA*, *leuB*, *leuC* and *leuD* (Keseler *et al.*, 2011). The *leuB* gene product codes for the enzyme 3-isopropylmalate dehydrogenase which catalyses the conversion of 3-isopropylmalate to 2-isopropyl-3-oxosuccinate (Fig. 2.8). The absence of this gene renders the strain auxotrophic for leucine when grown in minimal medium.

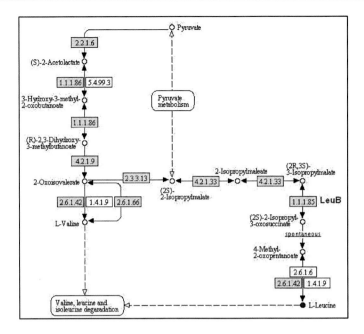

Fig. 2.8: Leucine biosynthesis pathway in *E. coli*. The *leuB* gene product 3-isopropylmalate dehydrogenase (EC 1.1.1.85) has been labelled. Image adapted from the online source KEGG (www.genome.jp/kegg) © 1995-2014 Kanehisa Laboratories.

2.6 The Keio collection

The auxotrophic host strain with the *leuB* or *tpiA* gene knockout, procured from CGSC (Coli Genetic Stock Center, USA) are strains from the Keio Collection, which is a systematic collection of *Escherichia coli* knockout strains in which each of the 3985 non-essential genes are represented by two independent deletion strains (Baba *et al.*, 2006) constructed using the Wanner's method of gene disruption (Datsenko & Wanner, 2000). The original parent strain used for the knock-outs is *E. coli* K12 BW25113 and the knocked out genes have been replaced in each case by a kanamycin resistance cassette during the course of construction of the strain (Fig. 2.9). This cassette is flanked by FLP recombinase sites and can therefore be removed by using *trans* functions from a plasmid such as pCP20 (CGSC, USA), which itself can be further cured by growth at specific temperatures, since it has a temperature sensitive origin of replication.

The homologous sequence carried as overhangs in the primers used for the recombination included a few bases of the target gene in the left and right flanking regions. As a result, on the upstream side, the start codon, and on the downstream side, the last six codons and the stop codon are retained on the genome (Baba *et. al.*, 2006). The reasoning behind this strategy was that the extreme ends of the coding sequence of a gene very frequently carried important regulatory and control functions in the expression of the upstream/downstream genes. The Keio knockout strain was desired to exhibit a phenotype that resulted exclusively from the loss of function of the target gene only and not due to effects on the flanking genes. The presence of the flanking codons of the target gene in the knockout genome brings into consideration the risk of reverse recombination between the genome and the cloned complementation gene on the plasmid, such that the knocked out gene may appear again in the genome and thus effectively remove the plasmid selection pressure.

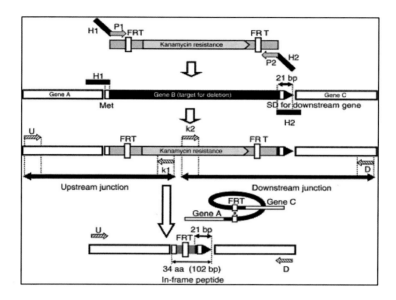

Fig. 2.9: Method of gene disruption followed to create the Keio knockout collection of *E. coli* mutants. Image adapted from Baba *et al.*, 2006. © 2006 EMBO and Nature Publishing Group.

2.7 Glycerol metabolic pathway

The central carbon metabolic pathway in *E. coli* could be exploited for the establishment of a catabolism-based auxotrophy complementation. The major routes for utilization of glycerol in *E. coli* are shown in Fig. 2.10. The gene *tpiA* represents the second target discussed in this work for the development of an antibiotic-free plasmid selection strategy. This gene codes for the enzyme triosephosphate isomerase and catalyses the conversion of dihydroxyacetone phosphate (DHAP) into glyceraldehyde-3-phosphate (G3P).

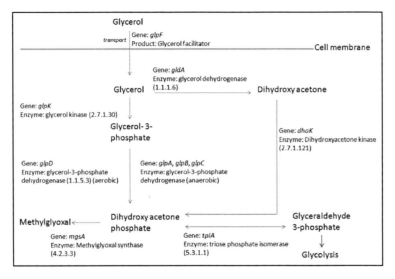

Fig. 2.10: Metabolic pathways involved in glycerol utilization in *E. coli*. Information taken from Hu & Wood, 2010 and from the online source EcoCyc (www.ecocyc.org; Keseler *et al.*, 2011).

When glycerol is transported through the cell membrane mediated by a diffusion facilitator protein (GlpF), two alternative pathways could be possible for the intracellular glycerol, depending on the growth conditions - a phosphorylation step followed by dehydrogenation or, an initial dehydrogenation step followed by phosphorylation. Of these, the former is the more predominant route (Keseler *et al.*, 2011). Accordingly, glycerol is converted to glycerol-3-phosphate by glycerol kinase (GlpK) which uses ATP as the phosphate donor. Under aerobic growth conditions, a homodimeric glycerol-3-phosphate dehydrogenase (GlpD) oxidizes glycerol-3-phosphate, whereas this step is catalyzed by a different heterotrimeric enzyme GlpABC under anaerobic conditions. The final metabolite that would result from any of the alternatives is DHAP. From this point, the flux can shuttle between DHAP and G3P, the latter being the key molecule to enter the glycolytic pathway. In the absence of the enzyme TpiA, DHAP can only be further metabolized to methylglyoxal. Moreover, the strategic position of *tpiA* on the metabolic map can be exploited to use a knockout host that allows some minimal growth on complex media but when transformed with an auxotrophy complementation plasmid, significant growth rate advantage for the plasmid-bearing cells.

2.8 Real time PCR

The analysis of the behaviour of stationary-phase or growth rate-dependent promoters in a continuous culture system is of scientific interest. In this work, promoter activity has been analysed by means of reverse transcription-quantitative real-time PCR of gene transcripts and fluorescence measurements from GFP as reporter protein. The focus is primarily upon the regulation of expression of the *kil* gene on plasmid p582 which is present under the control of the stationary phase promoter P_{fic}. The activation of this promoter for extracellular expression of the β-glucanase enzyme would be through change in the space velocity of the medium in a chemostat. Using promoter activity analysis, the molecular basis of the activation of P_{fic} could be verified. Keeping in line with the efforts at resolving ambiguity as detailed in the important paper on Minimum Information for Publication of Quantitative Real-Time PCR Experiments (MIQE) guidelines (Bustin *et al.*, 2009) quantitative real-time PCR of DNA would be called qPCR and reverse transcription-quantitative real-time PCR of RNA transcripts would be called RT-qPCR.

Reverse transcription-quantitative real-time PCR is a powerful method that allows the quantification of gene expression patterns in a wide variety of systems. The technique combines the specificity of mRNA analysis and the power of PCR amplification and real-time detection to yield useful quantitative estimations of the transcriptome in a cell and its variations under different conditions. The possibility of tracking the amplification of the target in real time differentiates this technique from conventional end-point PCR which does not allow accurate quantification of initial target concentration. RT-qPCR essentially involves isolating total RNA from a cell and using target-specific or random primers for reverse transcription of mRNA yielding RNA-cDNA hybrids or pure cDNA after degradation of RNA templates by RNaseH activity. These hybrids are then used as starting templates for a quantitative real-time PCR procedure that uses specific primers and DNA binding dyes to generate characteristic fluorescence profiles. By using a standard plot for the relation between known initial template concentration and its threshold fluorescence cycle (C_T), the initial quantity of specific mRNA transcript from an unknown sample could be estimated from its C_T value (Nolan *et al.*, 2006). By limiting the analysis to the kinetics of the PCR reaction during the exponential phase of the amplification, real-time PCR can provide estimations of the initial level of a target molecule whereas end-point PCR methods cannot (Leong *et al.*, 2007; Real-Time PCR Brochure, Qiagen Resource Center). Furthermore, the speed and sensitivity of this method makes it far more attractive than the earlier methods of analysis such as Northern blotting or RNase protection assay (Werbrouck *et al.*, 2007).

Fluorescent detection of amplification products is possible with either product-specific labelled-probes or by a non-specific dye. Labelled probes usually contain a fluorophore and a proximal quencher moiety. The probe and the primer bind to opposite ends of the target sequence and thus the extension of the primer releases the fluorophore allowing it to fluoresce. This technique enables further advancements like multiplex PCR wherein different targets are amplified and detected together in one reaction tube by using specific labelled-probes each with a different dye chemistry. The use of general DNA-binding dyes simplifies the procedure but entails the risk of extensive binding to non-specific amplicons or to primer-dimers which makes it very important to ensure primer specificity and efficiency. SYBR Green I (SG) is a highly sensitive cyanine dye that can bind with a high affinity to any dsDNA (Fig. 2.11) by intercalation and, at a dye to base pair ratio above 0.15, also by surface binding. The SG/dsDNA complex has a fluorescence emission maximum at 524 nm. Although, SG can bind to ssDNA which could cause high background fluorescence in real-time PCR assays, the resulting fluorescence intensity was found to be low (Zipper *et al.*, 2004). The intrinsic quenching of SG in its free state contrasts with the >1000 fold increase in fluorescence intensity when bound to dsDNA. The dye has been shown to be able to intercalate between base pairs, show electrostatic interactions and also intercalate at the DNA minor groove (Dragan *et al.*, 2012).

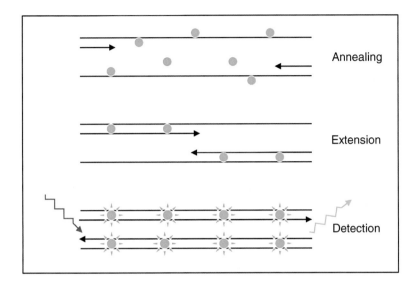

Fig. 2.11: Real-time PCR amplification and detection by the use of SYBR Green I dye. The dye binds double stranded DNA and allows fluorescence detection at the end of each amplification cycle. The resulting fluorescence is proportional to the amount of double stranded DNA at each cycle.

2.8.1 Theory behind real-time PCR amplification

The basic relationship representing PCR amplification is given by equation (1).

$$N_n = N_0 \cdot (1 + E_x)^n \qquad (1)$$

Where N_n is the number of amplicons of target x after n cycles, N_0 is the initial number of molecules and E_x is the efficiency of amplification of x with the given primers and at the given reaction conditions. Thus, at 100% efficiency, the value of E_x is 1 and the number of amplicons double every cycle. The equation also signifies that within the log-linear amplification phase, the amount of amplicons at any cycle is dependent on the initial amount of the template.

The use of a fluorescent DNA-binding dye such as SYBR Green allows defining a threshold level of fluorescence that signifies a certain level of amplification achieved. The cycle number at which the threshold fluorescence is crossed is called the threshold cycle (C_T). When C_T values are plotted against known amounts of templates, a standard curve can be generated with a slope m that helps in directly calculating the Efficiency (E_x) of amplification. The following figure (Fig. 2.12) shows this relationship in the form of a schematic graphical plot.

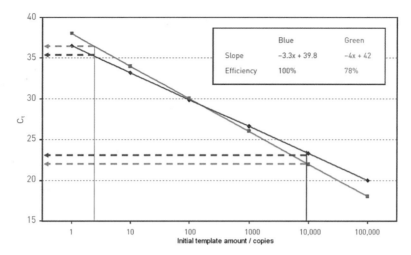

Fig. 2.12: Semi-logarithmic plot showing the relation between initial template amount and the threshold fluorescence cycle C_T for two targets amplified at unequal efficiencies to demonstrate the relationship between the efficiency and slope of the plot. Image adapted from Application Note – Real-Time PCR, Life Technologies. © 2011 Life Technologies Corporation.

Thus the efficiency of amplification can be calculated from equation (2),

$$E_x = 10^{(-1/m)} - 1 \qquad (2)$$

where m is the slope of the plot of C_T versus $\log_{10}$ of initial template concentration. When the amplification is 100% efficient, the plot has a negative slope of magnitude 3.322. A slope of less than -3.322 represents an efficiency less than 1.0 which signifies that the amplicons are not exactly doubling in number in each cycle (Real-Time PCR Brochure, Qiagen Resource Center). The physical significance of Fig. 2.12 is that, for a series of dilutions of the template, when the efficiency of amplification decreases it results in a steeper increase of the C_T resulting in delayed C_T at the low template concentration range. Inversely, the low efficiency also causes false low estimations of C_T at the high template concentration range.

2.8.2 Types of quantification

Quantification of the targets in a RT-qPCR may be absolute or relative. Absolute quantification aims to calculate the absolute amount of the target molecule in a reaction based on its position on a C_T plot that was generated from known standards. Relative quantification on the other hand expresses the ratio between the amounts of the target molecule and a suitable reference molecule whose concentration is not expected to vary widely. This ratio can then be compared among different samples, conditions or treatments thus giving an idea of differential gene expression.

Relative quantification is useful for molecular genetic analysis and can be performed when a calibrator condition is defined, such as the comparison of various kinds of tumour cells with a normal cell as calibrator. The calculation of the amounts of transcripts from each target gene requires the parallel measurement of a control gene for normalization which is mostly a housekeeping gene that is involved in basic cell function such as *gapA, rplM, gyrB* or the 16S rRNA gene. For eukaryotic systems, the genes coding for GAPDH, β-actin or ribosomal RNAs have been commonly used as housekeeping genes (Bustin, 2000). However, the term 'housekeeping gene' has of late become quite contentious since many experts have stressed the fact that seldom can a gene be assumed, without any validation, to be steady in its expression under all conditions (Gilsbach *et al.*, 2006; Bustin *et al.*, 2009). Therefore, there is a need to screen multiple candidate reference genes under the conditions of the particular assay in question. A useful algorithm called geNorm has been developed that helps in identifying the most stable reference gene through geometric averaging from among multiple candidates (Vandesompele *et al.*, 2002). An alternative method of normalisation is by spiking the samples with known quantities of an unrelated sequence of *in vitro* transcribed RNA (cRNA) that would serve as internal standard (Gilsbach *et al.*, 2006).

Once a stable reference gene has been identified, the difference in C_T (ΔC_T) between the reference and the target gene is checked over a wide concentration range. A constant ΔC_T would mean comparable amplification efficiencies between reference and target. After verifying that the amplification efficiencies are comparable, the change in C_T values between the target and the reference gene at the calibrator condition is first computed.

$$\Delta C_{T,\text{calibrator}} = C_{T,\text{target}} - C_{T,\text{reference}} \qquad (3)$$

The same parameter is then measured at the sample or condition to be compared.

$$\Delta C_{T,\text{sample}} = C_{T,\text{target}} - C_{T,\text{reference}} \qquad (4)$$

The difference in between these two ΔC_T values is denoted as $\Delta\Delta C_T$ for that particular sample or condition. Using the simple formula $\left((1 + E_x)^{-\Delta\Delta C_T}\right)$, the expression level of the target gene for a sample or condition normalized to an endogenous reference gene and relative to the calibrator sample or condition can be expressed (Livak & Schmittgen, 2001). Since in this work, the parameter under study is the specific growth rate of the culture, it would affect the expression of all genes globally and therefore it is difficult to expect a stable reference gene, making relative quantification unadvisable.

2.9 Green fluorescent protein for measurement of promoter activity

The concentration-dependent fluorescence of the green fluorescent protein from the jellyfish *Aequorea victoria* is quantifiable and offers a simple method of *in vivo* gene expression monitoring without the need for any additional chemical substrates and without any great stress to the host cell (Albano *et al.*, 1996). The primary structure of the wild type protein consists of 238 amino acids which when folded into the mature β-can conformation, forms a regular cylinder made of 11 β-sheets with an α-helix inside and a central essential Ser-Tyr-Gly tripeptide fluorophore region that is modified post-translationally (Fig. 2.13). The fluorophore is activated autocatalytically through a rapid cyclization between Ser 65 and Gly 67 followed by a rate-limiting oxygenation of the Tyr 66 (Yang *et al.*, 1996).

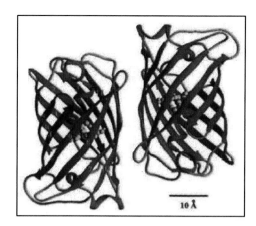

Fig. 2.13: GFP exists as a dimer in the crystal form with each monomer in a β-can formation made of 11 β-strands with an α-helix inside. The barrel also contains short α-chains on the top and bottom. The fluorophore is contained near the geometric centre. Image taken from Yang *et al.*, 1996. © 1996 Nature Publishing Group.

The use of a fast-folding variant allowed the development of a library of transcriptional fusions covering 75% of the promoters in *E. coli* for monitoring their activity during a diauxic shift experiment with high temporal resolution (Zaslaver *et al.*, 2006). Due to their extremely tolerant amino and carboxy termini, GFPs have also been used as fusion tags to monitor recombinant protein expression and solubilization while ideally still retaining the functions of both the target protein and the GFP tag (Sørensen & Mortensen, 2005).

The variant of *gfp* used in this study originates from Scholz *et al.* (2000) where *gfp+* was reported which combined the improved folding properties of the *gfpuv* variant and the chromophore mutations of GFPmut1 along with many silent mutations to adapt to the codon usage of *E. coli* resulting in 130 fold higher brightness and a sensitivity 320 times higher than the wild type, thus converting it into an ideal protein for detecting weak promoter activity (Scholz *et al.*, 2000). Although, the high stability of GFP is advantageous for certain objectives, it is necessary to follow the first derivative of fluorescence increase in order to monitor transient promoter activity (Zaslaver *et al.*, 2006). One important development would be to use unstable GFPs that offered excellent temporal resolution due to their rapid degradation and hence short half-life.

The principle is based on the natural *E. coli* system that uses the RNA molecule *ssrA* to tag partially synthesized polypeptides that have stalled due to incomplete or damaged mRNA. The addition of the amino acid sequence AANDENYALAA to the carboxy end, allows detection by specific cellular proteases that result in rapid degradation and turnover of the polypeptide (Andersen *et al.*, 1998). The house keeping ATP-dependent protease complexes ClpXP or ClpAP were found to be involved in this degradation. According to the model proposed, ClpX or ClpA have chaperone-like activity that recognizes, denatures and feeds substrate proteins into the active site chamber of the proteolytic component ClpP (Gottesman *et al.*, 1998). A variant of the degradation tag containing LVA as the last

3 amino acids gave the shortest half-life of about 40 min (Andersen *et al.*, 1998). The *gfp+* variant was thus improved for use in promoter activity analysis, by the addition of a LVA-type recognition tag to the carboxy end for protease degradation resulting in the gene *gfp+*-lva (Lässig, 2009). This fragment cloned onto a pET-24a(+) vector (Strain collection, Fermentation Engineering, Bielefeld University) is the starting point for the generation of pRS-ficGFP used in this study.

2.10 The stationary phase of bacterial growth

During batch growth, bacterial cultures go through distinct phases, of which the stationary phase marked by starvation and nutrient limitation is of particular relevance within the scope of this work. The global changes in regulation of gene expression upon entry into and during the stationary phase have been reviewed by Kolter *et al.* (1993) where the authors describe the entry into stationary phase to be the transition period beginning with the cessation of balanced growth and continuing up to the point when no further increase in cell number can be detected. The internal activation of the cell physiology is such that many functions induced during stationary phase are also induced when cells grow steadily at a low rate.

Bacterial RNA polymerase holoenyzme contains a primary or housekeeping sigma factor σ^{70} (RpoD) controlling the expression of genes during exponential growth. An alternative regulon controlled by sigma factor σ^S or σ^{38} (RpoS) is induced upon entry into stationary phase during nutrient-starved conditions (Kolter *et al.*, 1993). RpoS plays a major role in activating the expression of a multitude of genes involved in general stress response and survival including sugar metabolism, polyamine metabolism, transcription regulation, and nucleic acid metabolism and modification among others (Maciąg *et al.*, 2011). The major promoter for *rpoS* transcription called $rpoSp_1$ is present within the upstream structural gene *nlpD*. Control at the transcriptional level is provided by the signal molecule ppGpp playing a role in elongation (Lange *et al.*, 1995). With a length of 567 bp, the 5'-untranslated region of the *rpoS* mRNA transcript is one of the longest in *E. coli* and contains important sites for regulation by other RNAs and proteins that could promote mRNA stabilization and translation (Landini *et al.*, 2014). At the protein level, in exponentially growing cells, RssB can bind to and mark σ^S for proteolytic degradation by ClpXP proteases (Becker *et al.*, 2000). Yet another level of regulation is seen for the binding of the σ^S protein with the RNA polymerase core enzyme which is actively promoted by some proteins interacting directly with σ^S while yet others may do so indirectly by reducing the affinity of the core enzyme to σ^{70} (Landini *et al.*, 2014). Earlier, it was found that several σ^{70} promoters could also be recognized *in vitro* by the σ^S sigma factor while for many other promoters the recognition was specific for either of the two sigma factors. In fact, the -10 region of the promoters recognized by both the sigma factors is highly conserved (Tanaka *et al.*, 1993). Although both the sigma factors share many residues in their DNA-binding regions, apart from the influence of additional factors for promoter recognition, the selectivity of the RNA polymerase holoenzyme with either σ^{70} or σ^S towards their target promoters is influenced by different tolerances of these sigma factors to deviations from the common consensus at the -10 site and by other intrinsic features of the -10 and -35 promoter sites (Landini *et al.*, 2014).

2.11 Continuous cultivation

2.11.1 Theoretical background

The analysis of *E. coli* under continuous culture provides important insights into the physiological mechanisms and has potential for improvement of recombinant fermentation processes. In continuous culture, fresh medium is added continuously to a bioreactor while depleted medium, cells and products are continuously removed at rates such that the volume of the culture remains constant (Dunn *et al.*, 2003). Although there are a variety of strategies to operate continuous cultures, this work will be limited to discussing continuously operated stirred tank (bio-) reactors (CSTRs) which are characterized by a homogenous distribution of biomass and a neglect of any gradients in the operating variables. Two simple continuous culture devices referred to by their historical names are preferentially used for studying microbial processes: the chemostat and the turbidostat. The former refers to a constant chemical environment ('chemo'-'stat') in the reactor that is achieved by fixing the nutrient feed flow rate whereby the growth rate of the culture adjusts to the feed rate leading to a steady state for the operating variables. The turbidostat, on the other hand, is a controlled system in which the turbidity of the culture is maintained constant by adjusting the feed flow rate (Lee, 1992). The chemostat has been traditionally used in the laboratory as a tool to study the physiology and kinetics of microbial growth.

One of the earliest descriptions of the chemostat and its most basic principle was given by Novick & Szilard (1950). Referring to a Monod-type growth kinetics, they described that it is possible to maintain a microbial culture in a state of perpetual growth, at different rates all lower than the maximum (μ_{max}) by providing all but one nutrient in excess and using the concentration of the limiting nutrient as a controlling factor. By controlling the concentration of the limiting substrate and maintaining all physico-chemical conditions constant, the cells can be maintained at a constant state of growth and therefore physiological characteristics can be better studied than in a dynamically changing environment of a batch process. At steady state, the cell growth rate is determined by the rate of substrate feeding. By measuring the residual substrate concentrations at different space velocities, the growth kinetics of the particular strain can be established. In addition, other important parameters like yield and maintenance coefficients could be determined. The basic relation expressing growth rate as a function of limiting substrate concentration is given by the Monod function (Fig. 2.14) expressed in equation (5) (Monod, 1949).

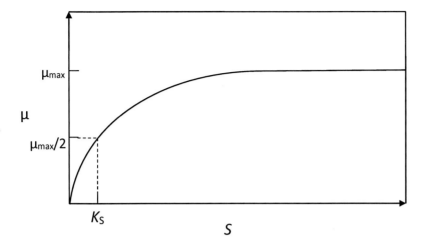

Fig. 2.14: Representation of the Monod function showing the relation between the specific growth rate (μ) of a bacterium as a function of the limiting substrate concentration (S) in the medium.

$$\mu = \frac{\mu_{max} \, S}{K_S + S} \tag{5}$$

where μ is the specific growth rate of the bacterial strain, S is the concentration of the limiting substrate, μ_{max} is the maximum specific growth rate and K_S is the Monod constant which represents the substrate concentration for which the strain grows at half the maximum rate.

In order to discuss the results, certain models for growth kinetics would be required which are simplified here through the following assumptions.

i) The microorganism follows a Monod-type of growth kinetics.

ii) The concentration of a single substrate limits the growth in a chemostat.

iii) The yield coefficient of biomass over substrate is either a constant overall yield coefficient or a case where some maintenance energy is deducted from the substrate resulting in an observed yield coefficient $Y_{X/S}$.

iv) Cell death rate or cell lysis within the system is not taken into account.

v) Mixing in the vessel is instantaneous and concentration gradients are absent.

vi) The chance of evolution of mutant cells with altered characteristics during long-term continuous cultivation is negligible.

2 Theory

Batch mode of operation

<u>Biomass balance</u>

In a closed system, feed rate $F = 0$ and therefore for the biomass growth,

$$\frac{dX}{dt} = r_X = \mu X \tag{6}$$

if cell death were to be considered, the expression would be,

$$\frac{dX}{dt} = \mu X - k_d X \tag{7}$$

where k_d is the cell death rate constant.

Rearranging eqn. (6) and integrating for limits X_0 to X corresponding to time 0 to t,

$$\ln\left(\frac{X}{X_0}\right) = \mu t \tag{8}$$

or

$$X = X_0\, e^{\mu t} \tag{9}$$

Eqn. (9) represents the relationship for biomass concentration in a batch reactor at time t, assuming exponential growth with a constant specific growth rate μ. X_0 is the initial biomass concentration.

<u>Substrate balance</u>

For the limiting substrate, a material balance can be written in the form of,

$$-\frac{dS}{dt} = r_S = q_S X \tag{10}$$

In eqn. (10), the specific substrate consumption rate q_S can be expressed in terms of the apparent biomass yield coefficient over substrate $Y_{X/S}$.

$$q_S = \frac{1}{Y_{X/S}} \cdot \mu \tag{11}$$

Continuous mode of operation

<u>Biomass balance</u>

Assuming that in a chemostat (Fig. 2.15), the flow rate of the feed inlet F_i is equal to the outlet flow rate F_o and is represented by F, that the feed is sterile and that there is no cell recycle,

Biomass accumulation = Biomass growth – biomass outflow – cell death

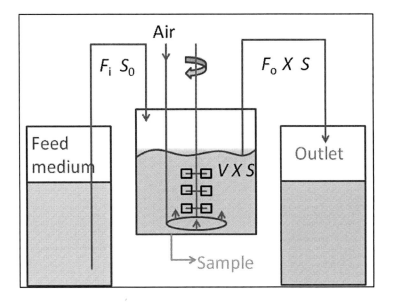

Fig. 2.15: Schematic representation of a continuous cultivation system

Ignoring cell death, the balance takes the form,

$$\frac{dX}{dt} = \mu X - \frac{F}{V} \cdot X \qquad (12)$$

At steady state, the accumulation term reduces to zero and therefore,

$$\mu = \frac{F}{V} = D \qquad (13)$$

which clearly signifies that the specific growth rate of the culture can be controlled by the space velocity D through the chemostat. The space velocity is also variably referred to as dilution rate in some sources.

The volumetric biomass productivity or output is given by

$$L_{V,X} = D \cdot X = \frac{F \cdot X}{V} \qquad (14)$$

Substrate balance

A substrate material balance for the chemostat can be written as

Substrate accumulation = Inlet feed − substrate utilization − substrate outflow

$$\frac{dS}{dt} = \frac{F}{V} \cdot S_0 - q_S \, X - \frac{F}{V} \cdot S \tag{15}$$

At steady state, substrate accumulation reduces to zero and therefore,

$$q_S \, X = D \cdot (S_0 - S) \tag{16}$$

The specific substrate utilization rate q_S can be expressed in terms of the apparent biomass yield coefficient to give,

$$\frac{1}{Y_{X/S}} \cdot \mu X = \mu \cdot (S_0 - S) \tag{17}$$

Eqn. (17) simplifies to give the expression relating the steady state biomass concentration and the initial substrate concentration in the feed through the definition of a constant yield coefficient.

$$X = Y_{X/S} \, (S_0 - S) \tag{18}$$

Rearranging equation (5) and substituting for μ at steady state, the substrate concentration can be expressed explicitly as

$$S = \frac{D \cdot K_S}{(\mu_{max} - D)} \tag{19}$$

From equation (19), the substrate fraction f given by S/S_0 can be derived as,

$$f = \frac{1}{s_{0S}} \cdot \frac{D_{rel}}{(1 - D_{rel})} \tag{20}$$

where s_{0S} is defined as a saturation parameter given by S_0/K_S. D_{rel} is the relative space velocity given by D/D_{max}.

The Monod equation (5) further allows the expression of the ratio of the space velocity at the critical point to the maximum specific growth rate. As the feed replaces the biomass in the vessel, the substrate fraction f reaches 1.

$$\frac{D_c}{\mu_{max}} = \frac{s_{0S}}{(1 + s_{0S})} \tag{21}$$

Case for maintenance energy

Maintenance effects are important to include in any structured model. Apart from increase in cell number and the formation of products, cells would require energy from substrate for maintenance of cellular structure, repair of damaged cell constituents and to maintain concentration gradients over cell membranes. Thus, the maintenance coefficient m_S is defined as that component of specific energy substrate consumption rate that does not result in any net biomass synthesis (Enfors & Häggström, 2000). If maintenance energy is considered, this factor needs to be included in the substrate balance, which modifies eqn. (16) to,

$$D \cdot (S_0 - S) = q_S X + m_S X \tag{22}$$

Expressing S in terms of the substrate fraction f and since maintenance is explicit, expressing q_S in terms of the true biomass yield coefficient $Y'_{X/S}$ eqn. (22) would expand to,

$$D \cdot S_0 (1 - f) = \left(\frac{1}{Y'_{X/S}} \cdot \mu_{max} \frac{S_0 f}{(K_S + S_0 f)} + m_S \right) \cdot X \tag{23}$$

Substituting with D_{rel} the equation can be solved for the biomass fraction Φ as,

$$\Phi = \left(\frac{1 - f}{D_{rel} + \Phi_S} \right) \cdot D_{rel} \tag{24}$$

where the maintenance parameter Φ_S is given by,

$$\Phi_S = \frac{m_S \cdot Y'_{X/S}}{\mu_{max}} \tag{25}$$

The dimensionless Damköhler number of the first kind for zero order kinetics ($Da_{I,0}$) is defined as the ratio of the amount of maximum converted educt to the amount of educt provided. It is inversely correlated to the relative space velocity as given by,

$$D_{rel} = \frac{1}{Da_{I,0}} \tag{26}$$

When no cell recycling from the outlet stream is employed, the chemostat is simply controlled by two main parameters – the space velocity (D) and the inlet feed substrate concentration (S_0) (Enfors & Häggström, 2000). While the former determines the specific growth rate (μ) the latter determines the maximum biomass concentration (X).

Thus, at steady state, the specific growth rate of the cells equals the space velocity of the medium through the reactor. In other words, as long as the system is being operated away from the critical space velocity, the cells sustain themselves in the chemostat by growing at a rate corresponding to the substrate concentration within the reactor. The equations provide the basis for examining the relationships between substrate feed rate, limiting substrate and biomass concentration and yield coefficient in a chemostat. The use of structured and segregated models further improve precision but also complexity to the estimation by including for example maintenance effects (m_S) and cell death rate constant (k_d) respectively.

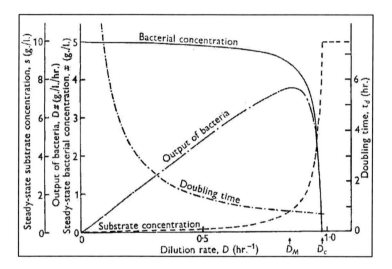

Fig. 2.16: Relation between steady-state biomass and substrate concentrations, biomass productivity, doubling time (t_d) and the space velocity (D) through the chemostat. Image adapted from Herbert *et al.*, (1956).

As seen in the schematic diagram in Fig. 2.16, for a constant volume V, increase in substrate inlet feed rate will increase the space velocity through the reactor. At steady state, the biomass and substrate in the chemostat (and outlet) reach constant concentrations. For low dilution rates, the steady state biomass concentration for a given dilution rate is determined by the feed substrate concentration. If the strain has a low K_S value, it is able to reach maximum specific growth rates already at relatively low substrate concentrations and hence would be able to maintain at high steady state biomass concentrations over a long range of space velocities. At the point where the strain's maximum specific growth rate is equal to the space velocity, growth would not be able to keep up with any further increase in feed rate and therefore the cells would be unable to sustain against the outflow. The biomass concentration falls to zero, the reactor substrate concentration eventually equals the feed substrate concentration and the system is said to have been washed out. As a result, the biomass productivity curve (referred to as Output of bacteria in Fig. 2.16) falls down too. In Fig. 2.17, it is shown that for a given substrate feed concentration, a gradual or early decline of biomass concentration with increasing space velocity, rather than a sharp fall near the critical point, is due to high K_S and indicates first-order growth kinetics. The following plot (Fig. 2.18) shows the effect of this phenomenon on the relative volumetric biomass productivity.

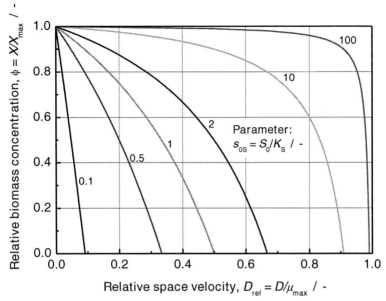

Fig. 2.17: Relation between the relative biomass concentration Φ and relative space velocity D_{rel} in a chemostat when the value of the saturation parameter is varied. Source: Flaschel, Lecture materials, Master course: Molecular Biotechnology, Bielefeld University.

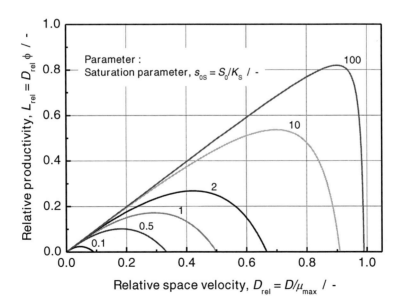

Fig. 2.18: Relation between the relative volumetric biomass productivity L_{rel} and the relative space velocity D_{rel} in a chemostat when the value of the saturation parameter is varied. Source: Flaschel, Lecture materials, Master course: Molecular Biotechnology, Bielefeld University.

For a given chemostat operating at steady state, there exists an optimal space velocity where the biomass productivity reaches a maximum. This is given by (Herbert *et al.*, 1956),

$$D_{opt} = \mu_{max} \cdot \left(1 - \sqrt{\frac{K_S}{(K_S + S_0)}} \right) \tag{27}$$

Non-ideal first-order growth rate kinetics

At high substrate concentrations, the specific growth rate of the cells is virtually independent of the substrate concentration. This is seen as the portion of the curve with $\mu = \mu_{max}$ in the Monod curve in Fig. 2.14 and is referred to as zero-order kinetics. The lower the value of K_S, the longer the portion of the Monod curve with maximum specific growth rate and therefore indicates the affinity of the cells for that particular medium. For a continuous cultivation, this would correspond to a wider range of substrate concentrations in which the cells can already grow at maximum rate and hence reach the steady-state biomass concentration according to the yield coefficient. However, this as well as the biomass vs. space velocity curve shown in Fig. 2.16 is an ideal case. Unfortunately often, a first-order kinetics is observed whereby, the specific growth rate is linearly proportional to the limiting substrate concentration. This corresponds to the region where S is less than K_S in the Monod curve which reduces the Monod equation (5) to

$$\mu = k' \cdot S \tag{28}$$

where $k' = \mu_{max}/K_S$. For a continuous cultivation at steady state, it can be derived from equation (28) that

$$S = K_S \cdot D_{rel} \tag{29}$$

and therefore,

$$f = \frac{D_{rel}}{S_{0S}} = D_{rel,1} = \frac{1}{Da_{I,1}} \tag{30}$$

and

$$\Phi = 1 - \frac{D_{rel}}{S_{0S}} \tag{31}$$

where D_{rel} is the relative space velocity (D/μ_{max}), f is the residual substrate fraction in the reactor (S/S_0) and s_{0S} is defined as a saturation parameter given by S_0/K_S. $Da_{I,1}$ is the dimensionless Damköhler number of the first kind for first order kinetics. These relationships are valid for the range $D_{rel} \leq s_{0S}$ and hence by defining the ratio D_{rel} / s_{0S} as the relative space velocity parameter $D_{rel,1}$ and plotting for various values of the maintenance parameter Φ_S, one can get the modified form of the ideal case of Fig. 2.16 in the form of Fig. 2.19 which is represented by

$$\Phi = \left(\frac{1 - D_{rel,1}}{D_{rel,1} + \Phi_S}\right) \cdot D_{rel,1} \tag{32}$$

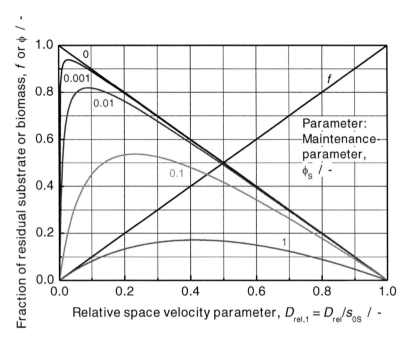

Fig. 2.19: Relation between fraction of biomass or residual substrate and the relative space velocity parameter $D_{rel,1}$ for a chemostat growing with first-order kinetics with respect to the limiting substrate shown for different values of the maintenance parameter. Source: Flaschel, Lecture materials, Bachelor course: Molecular Biotechnology, Bielefeld University.

Fig. 2.19 shows how the maintenance parameter affects only the biomass curve and not the substrate curve. Similarly, the relative volumetric biomass productivity is affected by the maintenance parameter as depicted in Fig. 2.20. It is interesting note that the maintenance parameter has no influence on the washout point but rather affects only the height of the curve. Both the washout point and the height of the curve would be controlled by the parameter s_{0S}.

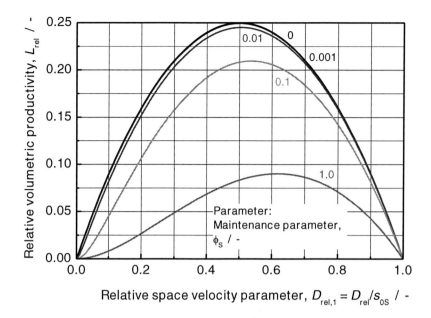

Fig. 2.20: Relation between relative volumetric biomass productivity L_{rel} given by $(D_{rel} \cdot \Phi)$ and the relative space velocity parameter $D_{rel,1}$ for a chemostat growing with first-order kinetics with respect to the limiting substrate shown for different values of the maintenance parameter. Source: Flaschel, Lecture materials, Bachelor course: Molecular Biotechnology, Bielefeld University.

The concept of maintenance requires two different biomass yield coefficients to be considered. The observed or apparent yield coefficient $Y_{X/S}$ and the true yield coefficient $Y'_{X/S}$ related to each other through the maintenance coefficient as shown in equation (33).

$$\frac{1}{Y_{X/S}} = \frac{1}{Y'_{X/S}} + m_S \cdot \frac{1}{\mu} \tag{33}$$

A plot of $1/Y_{X/S}$ versus $1/\mu$ gives a line with the slope equal to m_S and a y-intercept equal to $1/Y'_{X/S}$ which are considered to be constants. Thus, at low specific growth rates, which occur during low substrate concentrations, maintenance requirements still have to be met and therefore the cells experience a competition of growth versus maintenance. The equation shows the result of this phenomenon as a direct reduction in the observed yield coefficient.

The following 3 aspects to continuous cultivation played a crucial role within the scope of this work:

a) the expression of the stationary-phase P_{fic} promoter during low growth rate in continuous culture,

b) the stable operation of the system and avoidance of 'quasi-lysis' (Van der Wal *et al.*, 1995a) and

c) determination of the optimal space velocity for maximum extracellular volumetric productivity of the target enzyme. Only when these aspects could be initialized and studied, the next major step of incorporating an antibiotic-free plasmid selection strategy based on auxotrophic complementation could be implemented.

Cascade of CSTRs

For a system of two CSTRs connected in series (see Fig. 2.21) with the growth following a first order kinetics, the biomass concentration in the first vessel could be represented as the difference between the maximum and the hypothetical biomass possible from the residual substrate concentration as given in eqn. (34).

$$X_1 = X_{max} - Y_{X/S} \cdot \frac{D_N}{k'} \tag{34}$$

where the space velocity has been expressed so as to represent that for a cascade of multiple vessels. A plot of X versus D_N results in a line with a slope $(-Y_{X/S}/k')$, an X-intercept μ_0 corresponding to the specific growth rate at initial substrate concentration and a y-intercept at X_{max}.

The true biomass yield coefficient is the ratio between the theoretical maximum biomass concentration and the amount of initial substrate concentration.

$$Y'_{X/S} = \frac{X'_{max}}{S_0} \tag{35}$$

The biomass material balance for the second vessel can be represented as,

Biomass accumulation = biomass inflow + biomass growth – biomass outflow – cell death

Ignoring cell death and at steady state with a flow F, the above relation would be expressed mathematically as,

$$FX_2 - FX_1 = V.k'S_2 \cdot X_2 \tag{36}$$

where the subscripts 1 and 2 represent the individual chemostat vessels. From Eqn. (36) the biomass concentration in the second vessel could be expressed as,

$$X_2 = X_{max} - Y_{X/S} \cdot S_2 \tag{37}$$

Combining equations (36) and (37) and expressing S_2 in terms of X_2, a quadratic equation of the form $(ax^2+bx+c = 0)$ could be derived for the biomass concentration,

$$X_2{}^2 + \left(Y_{X/S} \cdot \frac{D_N}{k'} - X_{max}\right)X_2 - Y_{X/S} \cdot \frac{D_N}{k'} \cdot X_1 = 0 \tag{38}$$

Eqn. (38) could be converted to express the biomass fraction Φ and further generalized for the ith vessel, resulting in the solution for the biomass fraction given by the pair of roots,

$$\Phi_i = -\frac{1}{2}(D_{rel,N} - 1) \pm \sqrt{\left(\frac{1}{4}(D_{rel,N} - 1)^2 + D_{rel,N} \cdot \Phi_{i-1}\right)} \tag{39}$$

where the relative space velocity $D_{rel,N}$ has been substituted as,

$$D_{rel,N} = \frac{D_N}{D_{max}} = \frac{D_N}{k' \cdot S_0} \tag{40}$$

Case for maintenance energy

When considering maintenance energy, the substrate material balance for the second vessel has to be written as,

Substrate inflow – substrate outflow = utilization for growth + utilization for maintenance

which is expressed mathematically in the form of eqn. (41)

$$D \cdot (S_1 - S_2) = \frac{1}{Y'_{X/S}} \cdot k' \, S_2 \cdot X_2 + m_S \cdot X_2 \tag{41}$$

The biomass material balance for the second vessel given in equation (36) could be modified into,

$$\frac{D_N}{k'} \cdot X_2 - \frac{D_N}{k'} \cdot X_1 = S_2 \cdot X_2 \tag{42}$$

Expressing in terms of the biomass fraction Φ and using the eqn. (40) the solution would be,

$$\Phi_2 = \frac{D_{rel,N}}{D_{rel,N} - f_2} \cdot \Phi_1 \tag{43}$$

Which can also be correlated with the general form in eqn. (39) applied for the second vessel. When the concentrations of biomass and substrate in the first vessel have been solved for, these can be used after combining equations (41) and (42) to derive a generalized quadratic equation for the substrate fraction in the ith vessel as,

$$f_i^2 - (D_{rel,N} + f_{i-1} + \Phi_{i-1})f_i + (D_{rel,N} \cdot f_{i-1} - \Phi_S \cdot \Phi_{i-1}) = 0 \qquad (44)$$

which would give the following pair of roots for f_i,

$$f_i = \frac{1}{2}(D_{rel,N} + f_{i-1} + \Phi_{i-1}) \pm \sqrt{\left(\frac{1}{4}(D_{rel,N} + f_{i-1} + \Phi_{i-1})\right)^2 - (D_{rel,N} \cdot f_{i-1} - \Phi_S \cdot \Phi_{i-1})} \qquad (45)$$

Combining equations (41) and (42) it is also possible to derive a quadratic equation for the biomass concentration in the second vessel in terms of the concentration in the first vessel as given in eqn. (46),

$$-\left(1 + \frac{m_S \cdot Y'_{X/S}}{D_N}\right)X_2^2 + (X_1)X_2 + \left(\frac{Y'_{X/S}}{k'} \cdot D_N \cdot X_1\right) = 0 \qquad (46)$$

The kinetics for a cascade of vessels would be presented later in the discussion of results from 2-stage chemostat experiments with the reference strain (Fig. 5.26).

2.11.2 Use of continuous cultivation

With the recent advances in –omics technologies, continuous culture methods have received renewed interest for generating homogeneous samples that are more meaningful for analysis and for delivering data that are reliable and reproducible (Bull, 2010). Samples from the controllable and defined physico-chemical environment of chemostats were found to be better suited for global transcriptomic, proteomic or metabolomic analyses than those from the dynamic environments of batch processes which contained secondary effects due to growth-rate changes (Hoskisson & Hobbs, 2005; Mashego *et al.*, 2007). Physiological artefacts caused by, for example growth rate variations due to a dynamic growth environment could be eliminated, and significant but subtle biological effects could now be unraveled with much more confidence for reproducibility. The importance of maintaining cells in a steady and controllable physiological state using a chemostat and its superiority over batch process for gaining meaningful gene expression data was shown in the yeast system *Saccharomyces cerevisiae* and applied to studying the effects of carbon and nitrogen limitation on the transcriptome (Hayes *et al.*, 2002).

The chemostat offers a possibility to achieve high productivity while maintaining physiologically defined steady-state conditions. However, it has until now found only limited applications in industry like for growth-associated products (eg. ethanol) or in wastewater treatment plants (Enfors & Häggström, 2000). One of the main reasons for avoidance of continuous systems for commercial recombinant protein production is regulatory – during quality control, tracking of a particular lot of the product to a particular batch is not possible. The probability and risk of contamination is ever-present and higher than in a batch or fed-batch process. Other than that, continuous cultivation of recombinant strains runs the risk of creating genetic instability or harmful mutations. One interesting application of continuous culture is in bioethanol production since the elimination of batch turn-around times would translate to a significant operational-cost reduction. Knockout strains of *E. coli* K12 with conditionally lethal deletions of lactate dehydrogenase (*ldh*) and pyruvate formate lyase (*pfl*) genes which were unable to grow anaerobically, were complemented with plasmids with recombinant pyruvate decarboxylase (*pdc*) and alcohol dehydrogenase (*adh*) genes from *Zymomonas mobilis* that conferred ethanologenicity but also simultaneously complemented the growth defect and gave a selection pressure under anaerobic conditions (Martin *et al.*, 2006). The aspect of loss of plasmids was also considered in that, plasmid stability was tested in antibiotic-free continuous culture on xylose and glucose for efficient ethanol production with conditions mimicking industrial production plants where contact with oxygen is not completely preventable.

Nevertheless, the possibility to alter growth rate and maintain at a desired value without altering the physico-chemical environment of the cells and further to change the physico-chemical environment as needed (Gilbert, 1985) makes the chemostat an important tool in the laboratory to study microbial physiology, growth and product formation. Shifting of metabolic flux from a high energy consumption state to an energy-efficient state favoring biomass formation at high space velocities in an aerobic glucose-limited continuous culture of *E. coli* K-12 strain TG1 was reported by Kayser *et al.* (2005). The chemostat offers an excellent possibility to study physiological changes as a result of experimental perturbations in cells growing under steady and defined conditions. Recently, the advantage of having a constant physico-chemical environment in a chemostat was utilized to create perturbations in the form of a glucose pulse and analyzing the transient physiological responses of *E. coli* to the shift between glucose-limited and glucose-excess conditions (Sunya *et al.*, 2012). Similarly, the adaptations observed as changes in the proteome and generation of mutants as a consequence of changing environmental conditions from glucose-excess batch to glucose-limited chemostat were studied. Interestingly, under glucose-limited conditions many periplasmic binding proteins belonging to other transport systems like maltose and galactose were found to be induced (Wick *et al.*, 2001), and in fact, this phenomenon was described earlier in a review to be part of a complex set of processes designed to improve the strain's scavenging ability under nutrient-limiting conditions (Kovárová-Kovar & Egli, 1998). In other industrially significant bacteria such as *Corynebacterium glutamicum*, which is important for the production of amino acids, the study of metabolic fluxes from glucose-limited chemostat cultures at various growth rates has helped to shed light on secondary glucose transport systems and a shift in the profile of anaplerotic pathway enzymes, that are crucial for amino acid synthesis rates and yields (Cocaign-Bousquet *et al.*, 1996). On the reverse side, the constant availability of all nutrients in excess and a single growth rate-controlling nutrient in limiting concentration also represents an artificial situation, which makes chemostat cultures quite unsuitable to study microbial kinetic relationships as they occur in the natural environment (Kovárová-Kovar & Egli, 1998). The possibility for strain improvement was demonstrated by Groeneveld *et al.*, (2009) using pH-controlled continuous cultivations that resulted in a fast growing strain of the yeast *Kluveromyces marxianis* with applications in single-cell protein production.

Overproduction of recombinant α-amylase from a *Bacillus* sp. and even extracellular expression by continuous culture of *E. coli* in an airlift fermenter was reported by Alexander *et al.*, (1989). Continuous cultures of *E. coli* have been studied for the production of recombinant human interferon-gamma as inclusion bodies (Vaiphei *et al.*, 2009), however, the nature of variable media and feeding pattern employed along with the IPTG induction of the T7-based system resulted in inevitable change of cell characteristics post-induction. Although the specific productivity was shown to increase with the specific growth rate, all measured product values were represented as values at hours post-induction and therefore, the long-term stability and productivity of such a system is not known. More recently, the effect of process temperature on the periplasmic expression of recombinant antibody fragments (Fab) in continuous cultures of medium cell densities was studied, albeit with decrease in product formation over time probably as a result of plasmid loss due to selection against plasmid-bearing cells (Rodríguez-Carmona *et al.*, 2012). Though the product leakage from periplasm to the extracellular space was shown to be improved at lower temperatures, the exact mechanism of the leakage was not known. Continuous culture is being increasingly used in industrial practice if the production strains exhibit sufficient biological stability for long-term cultivation processes. This is particularly true for slow-growing microorganisms wherein techniques of biomass retention such as perfusion systems need to be applied. This would additionally help in extending the range of feed rates that could be operated, minimize residence time for products to preserve quality and help establish integrated production processes (Warikoo *et al.*, 2012; Ahn *et al.*, 2008).

2.11.3 Two-stage chemostat

A cascade of stirred tank reactors was evaluated in this work, to find out if it offered an advantage over a simple CSTR, particularly considering the age distribution in the cell population that might favour activation of the stationary phase P_{fic} promoter. The 2-stage chemostat consists of a feed reservoir and reactors R1 and R2 and an outlet for the grown culture broth. The outlet from R1 containing cells, products and residual substrate flows into R2 as feed. Both reactors contain cultures of equal working volume, can be sampled independently and the flow rate through the system is maintained equal (Fig. 2.21). Such a cascade would make it possible to realize and study different conditions at each stage which would not be possible with a single stage chemostat (Bull, 2010).

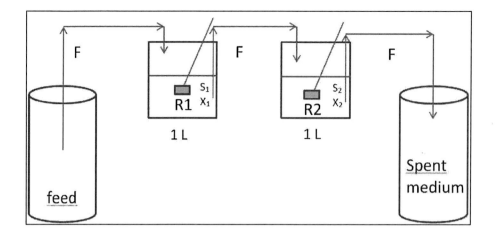

Fig. 2.21: Schematic representation of a 2-stage chemostat consisting of reactors R1 and R2 placed in series.

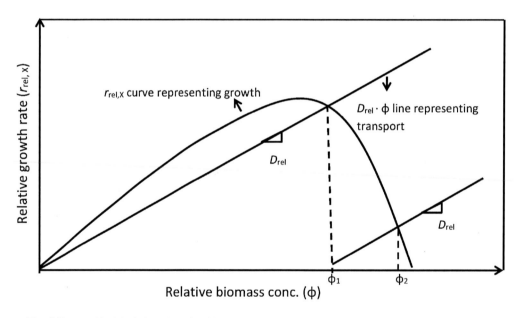

Fig. 2.22: Graphical depiction of relative biomass growth rate versus relative biomass concentration in a two-stage process.

For a general case, assuming the relative biomass growth rate in the reactor is given by $r_{rel,X}$ and the relative specific biomass growth rate by μ_{rel}, the relative biomass concentration is given by ϕ (X/X_{max}) so that $\phi = (1-f)$ where f is the residual substrate fraction (assuming negligible maintenance effects).

Considering the feed to R1 to be sterile,

$$r_{rel,X} = \frac{r}{r_{max}} = \mu_{rel} \cdot \Phi \tag{47}$$

In the plot in Fig. 2.22 the point of intersection of the growth curve and the transport line represents a steady state where $\mu_{rel} = D_{rel}$. At this point the corresponding biomass fraction in the first vessel is ϕ_1. In the next vessel, the culture enters with a handicap since it 'brings in' ϕ_1 fraction of cells and increases the biomass apart from a small contribution from the growth curve. The transport term for this stage is the same since the feed flow rate is the same through both the vessels but the growth is different due to the changed residual substrate concentration. The result is a parallel transport line starting from ϕ_1 and meeting the growth curve at a higher biomass fraction ϕ_2 but lower relative growth rate.

Use of a cascade of two CSTRs in steady-state cultivation of hybridomas for monoclonal antibody production has allowed the kinetic study of conditions in the second reactor that are difficult to achieve with a single reactor such as very low growth rates, defined death rates and cell lysis (Bakker *et al.* 1996). Physiological effects resulting from toxic influences on a 2-stage whole-cell biotransformation process using *Saccharomyces cerevisiae* were studied by applying the precursor (stressor) only to the 2nd stage while unaffected cells were continuously fed from the 1st stage. This allowed the system to be operated at identical steady states in both stages, followed by biotransformation experiment in the 2nd stage and analysis of the changes to cell physiology without the risk of washout (Hortsch *et al.*, 2008). A similar advantage over single stage chemostat was seen in the continuous production of recombinant β-lactamase with a *E. coli* K-12 strain. Here, the aim was to separate growth and production in order to achieve plasmid stability and prevention of appearance of lac⁻ mutants in the first stage and induction of recombinant gene using IPTG in the 2nd stage. The continuous supply of unaltered cells from the first stage acted as a backup for cells affected by the induction (Fu *et al.*, 1993).

3 Objectives

The development of the BRP strategy greatly improved the efficiency of extracellular recombinant protein production primarily due to the clever choice of promoter controlling the expression of the *kil* gene. The weak stationary-phase promoter of the *fic* gene when used to control *kil*, helped to achieve secretion at the end of a batch process and also the necessary moderate expression to prevent cell lysis at low growth rate during fed-batch. Thus one of the main objectives in this work was to setup a continuous culture process for the production of the model product β-glucanase and to ensure the activity of the 'stationary-phase' P_{fic} promoter by careful control of the specific growth rate by controlling the space velocity. Thereby, it was essential to study if a stable operation of the chemostat was possible at all with a background expression of the BRP which could result in washout of the culture due to quasi-lysis. In the case of success, this could lead to studying the kinetic relationships between biomass and product expression at different growth rates in the chemostat to find the optimal space velocity for maximum β-glucanase productivity.

The construct p582 had been shown to be an efficient plasmid for extracellular recombinant β-glucanase expression in the reference strain *E. coli* JM109 in batch and fed-batch conditions before demonstrating its functionality for the first time in continuous cultivation. Simultaneously, the segregational instability of this plasmid in the absence of antibiotic selection pressure was also demonstrated, thus setting up the basis for the genetic modifications of p582 through the construction of an auxotrophic complementation system comprising of the cloned genes *leuB* or *tpiA* on the target plasmid p582 and the use of auxotrophic knockout mutants from the Keio collection. The growth and product expression characteristics of the alternative systems were to be studied in shake-flask and batch modes before studying the continuous mode of operation. Furthermore, the efficiency of the alternative selection principle had to be compared to antibiotics-based systems in long-term continuous cultivation. Additional genetic modifications in the complementation plasmid and the auxotrophic host were to be tested for any improvements in productivity of the alternative system. Thus, the objective would be a system representing the combination of continuous culture with *E. coli*, extracellular product expression, and plasmid stability in an antibiotic-free process using an optimized auxotrophic complementation system.

In addition, the expression of the *kil* gene was to be followed in continuous culture together with genes assumed to be involved in the regulation of the P_{fic} promoter that controlled the expression of the *kil* gene. Therefore, the activity of the P_{fic} promoter was to be studied at different specific growth rates in a chemostat. To add more information to this analysis, it would also be essential to track the change in relative plasmid abundance per genomic DNA equivalent as a function of the specific growth rate. These experiments should offer insight into the activation of the growth-phase regulated promoter P_{fic} under different space velocities which would be a significant step towards optimization of the continuous culture process.

4 Materials and Methods

Materials used in the various experiments are noted within the description of the experimental procedures. All strains, plasmids, growth media, compositions of solutions as well as other commonly used chemicals and instruments along with their suppliers are listed in the following tables. Commercial kits and specialty chemical or biochemical preparations for molecular genetics applications are mentioned within the text in the description of the procedures.

4.1 Strains and plasmids

The *E. coli* strains used in this study, their sources, and properties are listed under Table 4.1.

Table 4.1: List of *E. coli* strains used in this study.

Strain	Source	Genotype	Remarks
JM109	Promega GmbH, Germany	*endA1, recA1, gyrA96, thi, hsdR17* (r_k^-, m_k^+), *relA1, supE44,* Δ(*lac-proAB*), [F′, *traD36, proAB, lacI^{q}Z*ΔM15]	Reference host strain for β-glucanase expression; Cloning strain
DH5α	Invitrogen, Germany	F$^-$, Φ80*lacZ*ΔM15, Δ(*lacZYA-argF*), U169, *recA1, endA1, hsdR17* (r_k^-, m_k^+), *phoA, supE44, thi-1, gyrA96, relA1,* λ$^-$	Cloning strain
K12 MG1655	Leibniz Institute DSMZ, Germany	F$^-$, λ$^-$, *rph-1*	Strain for genomic DNA isolation
Top10	Invitrogen, Germany	F$^-$, *mcrA,* Δ(*mrr-hsdRMS-mcrBC*), Φ80*lacZ*ΔM15, Δ*lacX74, recA1, araD139,* Δ(*ara leu*)*7697, galU, galK, rpsL* (*str^R*), *endA1, nupG*	Cloning strain
BW23474	CGSC, Yale University, USA	F$^-$, Δ(*argF-lac*)*169,* Δ*uidA4::pir-116, recA1, rpoS396*(Am), *endA9* (del-ins)::FRT, *rph-1, hsdR514, rob-1, creC510*	Strain with marker coding for Pir initiator protein
BW25113	CGSC, Yale University, USA	F$^-$, Δ(*araD-araB*)*567,* Δ*lacZ4787* (::*rrnB-3*), λ$^-$, *rph-1,* Δ(*rhaD-rhaB*)*568, hsdR514*	Parent strain for the Keio collection
JW5807-2	CGSC, Yale University, USA	F$^-$, Δ(*araD-araB*)*567,* Δ*leuB780::kan^R,* Δ*lacZ4787* (::*rrnB-3*), λ$^-$, *rph-1,* Δ(*rhaD-rhaB*)*568, hsdR514*	*leuB* knockout strain from Keio collection
JW3890-2	CGSC, Yale University, USA	F$^-$, Δ(*araD-araB*)*567,* Δ*tpiA778::kan^R,* Δ*lacZ4787* (::*rrnB-3*), λ$^-$, *rph-1,* Δ(*rhaD-rhaB*)*568, hsdR514*	*tpiA* knockout strain from Keio collection
KRX	Promega GmbH, Germany	[F′, *traD36,* Δ*ompP, proA$^+$B$^+$, lacI^q,* Δ(*lacZ*)M15] Δ*ompT, endA1, recA1, gyrA96* (*nal^R*), *thi-1, hsdR17* (r_k^-, m_k^+), e14$^-$ (*mcrA$^-$*), *relA1, supE44,* Δ(*lac-proAB*), Δ(*rhaBAD*)::T7RNAP	Strain for cloning and protein expression

HS996	Gene Bridges GmbH, Germany	F⁻, *mcrA*, Δ(*mrr-hsdRMS-mcrBC*), Φ80*lacZ*ΔM15, Δ*lacX74*, *recA1*, *araD139*, Δ(*ara-leu*)7697, *galU*, *galK*, *rpsL* (*str*^{R}), *endA1*, *nupG*, *fhuA*::IS2	Control strain for gene deletion
BW6	This work	F⁻, Δ(*araD-araB*)567, Δ*tpiA*::*kan*^{R}, Δ*lacZ4787* (::*rrnB-3*), λ⁻, *rph-1*, Δ(*rhaD-rhaB*)568, *hsdR514*	Wide *tpiA* gene deletion in the Keio parent strain
JM1	This work	*endA1*, *recA1*, *gyrA96*, *thi*, *hsdR17* (r$_k^-$, m$_k^+$), *relA1*, *supE44*, Δ(*lac-proAB*), Δ*tpiA*::*kan*^{R}, [F′, *traD36*, *proAB*, *laqI*^{q}*Z*ΔM15]	Wide *tpiA* gene deletion in the control strain JM109
JM2	This work	*endA1*, *recA1*, *gyrA96*, *thi*, *hsdR17* (r$_k^-$, m$_k^+$), *relA1*, *supE44*, Δ(*lac-proAB*), Δ*tpiA*::*kan*^{R}, [F′, *traD36*, *proAB*, *laqI*^{q}*Z*ΔM15]	Wide *tpiA* gene deletion in the control strain JM109

4.1.1 Description of important features from genotype of reference strain *E. coli* JM109

The marker *endA1* refers to a modification of endonuclease I that renders the strain suitable for the preparation of good quality plasmids. The *recA1* modification minimizes unwanted recombination of cloned DNA into the host genome. The modification *thi-1* refers to thiamine auxotrophy necessitating the supplementation of thiamine for growth in minimal medium. This is a safety feature to prevent the survival of accidentally released laboratory strains into the environment. The marker *relA1* refers to an insertion into the structural gene *relA* causing a mutation in ppGpp synthetase I. The marker *hsdR17* (r$_k^-$, m$_k^+$) refers to the mutation in the endogenous EcoKI restriction-modification system that disables restriction of unmethylated DNA by the restriction endonuclease subunit and thus ensures safe transfer of foreign DNA into the cell, whereas the methyltransferase subunit is active so that methylation of foreign DNA is still possible.

4.1.2 Description of important features of the strain *E. coli* JW5807-2

All mutations other than the *leuB* deletion were present already on the Keio parent strain *E. coli* K12 BW25113. The marker F⁻ refers to the absence of the F plasmid, the presence of which enables pili formation. Δ(*araD-araB*)567 refers to disruption in the arabinose metabolism by a deletion starting from 25 bp upstream of the gene *araB* through *araA* and 8 bp into *araD* (CGSC, Yale University). Δ(*rhaD-rhaB*)568 refers to disruption in the rhamnose metabolic genes. These modifications have no bearing on the experiments performed with the strain in this study. The leucine auxotrophy caused by the replacement of the *leuB* gene with the kanamycin resistance gene is represented by the marker Δ*leuB780*::*kan*^{R}. Due to modification of the DNA restriction-methylation system (*hsdR514*), cloning is possible without cleavage of transformed DNA by endogenous restriction endonucleases. The marker Δ*lacZ4787* (::*rrnB-3*) refers to the insertion of 4 tandem copies of the *rrnB* transcriptional terminator into the region from near the SacII site of *lacZ* through the promoter (CGSC, Yale University). The deletion of the lambda lysogen is represented by the marker λ^-, and *rph-1* refers to a frameshift mutation of 1 bp of the RNase PH gene. The Keio strains have been designed with the general pattern where the deletion extends from the region between the start codon and the last six codons before the stop codon of the target gene, thus leaving these codons intact (Baba *et al.*, 2006). As an exceptional case, the choice of homologous sequences for the primers in the case of *E. coli* JW5807-2 for the *leuB* knockout has resulted in the deletion of the start codon as well.

The strain *E. coli* JW3890-2 is similar to the strain JW5807-2 with the obvious exception that the *tpiA* gene has been knocked out instead of the *leuB* gene.

The plasmids used or created during this project are listed under Table 4.2.

Table 4.2: List of plasmids used in this study.

Plasmid	Source; reference	Important markers	Remarks
p582	Strain collection, Fermentation Engineering, Bielefeld University; Beshay *et al.*, 2009	ori (pUC19), *bla* (*amp*^R), *npt* (*kan*^R), P_{CP7}-*bgl*, T7 terminator, P_{fic}-*kil*, MCS	Reference plasmid for β-glucanase expression
p55	Miksch; Sommer *et al.*, 2009	ori (pUC19), P_{bgl}, MCS, *bla* (*amp*^R)	Derived by inserting a 0.5 kb fragment with promoter P_{bgl} into MCS of pUC19
p55-leuB	This work;	ori (pUC19), P_{bgl}-*leuB*, *bla* (*amp*^R)	Derived by cloning *leuB* fragment under P_{bgl} promoter in p55
pFC2	This work;	ori (pUC19), *bla* (*amp*^R), *npt* (*kan*^R), P_{CP7}-*bgl*, T7 terminator, P_{bgl}-*leuB*, P_{fic}-*kil*	Derived by cloning fragment P_{bgl}-*leuB* into p582
pFKN	This work; Knüttgen, 2013	ori (pUC19), *bla* (*amp*^R), P_{CP7}-*bgl*, T7 terminator, P_{bgl}-*leuB*, P_{fic}-*kil*	Derived by removing *kan*^R from pFC2 through double digest with Van91I and PspXI
pJET-tpiA	Strain collection, Fermentation Engineering, Bielefeld University	ori (pMB1), *bla* (*amp*^R), *eco47IR*, P_{lacUV5}, P_{T7}, MCS, P_1/P_2-*tpiA*	*tpiA* expression cassette with natural promoter and terminator cloned blunt into pJET backbone
pFC1	This work;	ori (pUC19), *bla* (*amp*^R), *npt* (*kan*^R), P_{CP7}-*bgl*, T7 terminator, P_1/P_2-*tpiA*, P_{fic}-*kil*	*tpiA* expression cassette with natural promoter and terminator cloned into p582 in forward orientation

pFC4	This work;	ori (pUC19), *bla* (*amp*^{R}), *npt* (*kan*^{R}), P$_{CP7}$-*bgl*, T7 terminator, P$_1$/P$_2$-*tpiA*, P$_{fic}$-*kil*	*tpiA* expression cassette with natural promoter and terminator cloned into p582 in reverse orientation
pJET-CP19tpiA	This work	ori (pMB1), *bla* (*amp*^{R}), *eco47IR*, P$_{lacUV5}$, P$_{T7}$, MCS, P$_{CP19}$-*tpiA*	Natural promoter of *tpiA* in pJET-tpiA replaced with weak promoter CP19
pJET-CP33tpiA	This work	ori (pMB1), *bla* (*amp*^{R}), *eco47IR*, P$_{lacUV5}$, P$_{T7}$, MCS, P$_{CP33}$-*tpiA*	Natural promoter of *tpiA* in pJET-tpiA replaced with weak promoter CP33
p582-CP19tpiA	This work	ori (pUC19), *bla* (*amp*^{R}), *npt* (*kan*^{R}), P$_{CP7}$-*bgl*, T7 terminator, P$_{CP19}$-*tpiA*, P$_{fic}$-*kil*	*tpiA* with promoter CP19 cloned into p582
p582-CP33tpiA	This work	ori (pUC19), *bla* (*amp*^{R}), *npt* (*kan*^{R}), P$_{CP7}$-*bgl*, T7 terminator, P$_{CP33}$-*tpiA*, P$_{fic}$-*kil*	*tpiA* with promoter CP33 cloned into p582
pRedET (tet)	Gene Bridges GmbH, Germany	oriR101 (pSC101), *repA*, *tet*^{R}, *araC*, P$_{BAD}$-*redγβα*, *recA*	Low copy number, temperature sensitive origin of replication, tetracycline resistance, phage λ recombination genes under arabinose-inducible promoter
pRedET (amp)	Gene Bridges GmbH, Germany	oriR101 (pSC101), *repA*, *amp*^{R}, *araC*, P$_{BAD}$-*redγβα*, *recA*	Low copy number, temperature sensitive origin of replication, ampicillin resistance, phage λ recombination genes under arabinose-inducible promoter
pFRT	Gene Bridges GmbH, Germany	ori (R6K), *npt* (*kan*^{R}), flanking FRT sites	Conditional origin of replication, selection cassette with kanamycin resistance gene
pBR322	Strain collection, Fermentation Engineering, Bielefeld University	ori (pMB1), *rop*, *amp*^{R}, *tet*^{R}, MCS	Cloning vector containing resistance genes against ampicillin and tetracycline
pJET-rpoS-frag.	This work	ori (pMB1), *bla* (*amp*^{R}), *eco47IR*, P$_{lacUV5}$, P$_{T7}$, MCS, *rpoS* fragment	Short fragment from *rpoS* cloned blunt into pJET backbone

pJET-rpoD-frag.	This work	ori (pMB1), *bla* (*amp*^{R}), *eco47IR*, P$_{lacUV5}$, P$_{T7}$, MCS, *rpoD* fragment	Short fragment from *rpoD* cloned blunt into pJET backbone
pJET-kil-frag.	This work	ori (pMB1), *bla* (*amp*^{R}), *eco47IR*, P$_{lacUV5}$, P$_{T7}$, MCS, *kil* fragment	Short fragment from *kil* cloned blunt into pJET backbone
pET-24a(+)-gfp+-lva	Strain collection, Fermentation Engineering, Bielefeld University	ori (pBR322), *rop*, P$_{T7}$-*gfp*+-lva, His-Tag, T7 terminator, *kan*^{R}, *lacI*	*gfp* gene with lva tag cloned into pET-24a(+) (Invitrogen) under T7 promoter
pRS-ficGFP	This work	ori (pBR322), *rop*, P$_{fic}$-*gfp*+-lva, His-Tag, T7 terminator, *kan*^{R}	T7 promoter of pET-24a(+)-gfp+-lva replaced with P$_{fic}$ promoter. *lacI* removed as well.

4.1.3 Information on the expression plasmid p582

The reference β-glucanase expression plasmid p582 has been described by Beshay *et al.* (2009) and represents the starting point for the experiments in this work. The plasmid map generated after complete sequencing is shown in the Appendix (Fig. 8.1). It contained a pUC19 backbone to which a *kil-Km* secretion cassette (Miksch *et al.*, 1997a) had been cloned. Plasmid pPhyt109 (Miksch *et al.*, 2002) was modified by exchanging the phytase expression cassette with a β-galactosidase expression cassette under the control of the constitutive synthetic promoter CP7 which had been categorized as a strong promoter (Jensen & Hammer, 1998). Thereafter, the β-galactosidase gene was replaced by the hybrid β-glucanase gene from pET20-bgl-his-sec (Beshay *et al.*, 2007b). The hybrid endo-(1-3,1-4)-β-glucanase gene H1 (*bgl*) contained codons for the 107 amino-terminal residues of the *B. amyloliquefaciens* β-glucanase and codons for the 107 carboxyl-terminal residues of the *B. macerans* β-glucanase. The amino terminal signal sequence of H1 for targeting to periplasm was the native *B. amyloliquefaciens* signal sequence. This particular hybrid gene combination was shown to exhibit a lower pH optimum and better thermostability (Borriss *et al.*, 1989). In the plasmid p582, the *bgl* gene further contained a His-tag at the carboxyl end to assist in downstream purification of the product. The *kil* gene present on this plasmid coded for the release protein for colicin E1 and was present under the control of the P$_{fic}$ promoter of the *fic* gene. The *fic* gene was thought to be involved in cell division (Kawamukai *et al.*, 1989) and its promoter has been shown to be weak in *E. coli* (Miksch & Dobrowolski, 1995) and recognized preferentially by the stationary-phase RNA polymerase sigma factor RpoS (Miksch & Dobrowolski 1995; Tanaka *et al.*, 1993). Furthermore, it was shown through various analyses that the −10 promoter site (TATACT) was essential for σ^S-based initiation of transcription from the P$_{fic}$ promoter (Hiratsu *et al.*, 1995). As mentioned earlier, the plasmid p582 carried the recombinant β-glucanase gene (*bgl*) with origins from the work by Borriss *et al.* (1989). However, upon analysis of the sequence of p582 and comparison to the sequence presented in Borriss *et al.* (1989), four mutations were discovered in the coding sequence, causing altogether three amino acid changes (Appendix, Fig. 8.6). Thus the hybrid *bgl* gene H1 originally described by Borriss *et al.* (1989) must have somehow through the years incorporated four mutations in its sequence and this could be an important aspect to note when referring to the source in future. The plasmid p582 used in this work is from the strain collection of the Fermentation Engineering group, Bielefeld University (kindly provided by Dr. Miksch).

The primers used in this work are listed below in Table 4.3. Underlined sequences refer to restriction sites or homologous bases in the 5'-overhangs.

Table 4.3: List of primers

Primer ID	Short description	Sequence (5'→ 3')
1	Fwd. primer *leuB* amplification	GCTCTAGATCAACACAACGAA AACAAC
2	Rev. primer *leuB* amplification	GGAATTCTTTACACCCCTTCTG CTAC
24	Fwd. primer for resistance gene deletion	ACTACGGCTACACTAGAAGG
25	Rev. primer for resistance gene deletion	GCGTCAGACCCCGTAGAAAG
5	Fwd. primer for *tpiA* fragment amplification from genome	TAAGCTGGCGCTATCTGAATCG
6	Rev. primer for *tpiA* fragment amplification from genome	GATGGTACGGCAGAGTGATAA C
7	Fwd. primer for *tpiA* fragment amplification from pJET-tpiA	GGAATTCTAAGCTGGCGCTATC TG
8	Rev. primer for *tpiA* fragment amplification from pJET-tpiA	GGAATTCGATGGTACGGCAGA GTG
19-1	Fwd. primer for replacing promoter on pJET-tpiA with CP19	GGGATCCGGGTTGATATAATA GTTAAGACCTGCTGCCCTGCGG GG
19-2	Rev. primer for replacing promoter on pJET-tpiA with CP19	TCCTGTCAAGAAAAACTAAGC GATGTTCAAGCGATTCAGATA GCG
33-1	Fwd. primer for replacing promoter on pJET-tpiA with CP33	AATTACTGCAGTGATATAATAG GTGAAGACCTGCTGCCCTGCG GGG
33-2	Rev. primer for replacing promoter on pJET-tpiA with CP33	GTATGTCAAGAATAAACTCCA ACATGTTCAAGCGATTCAGATA GCG

pJET 1.2 fwd	Forward sequencing primer from Thermo Scientific for sequencing of cloned inserts	CGACTCACTATAGGGAGAGCGGC
pJET 1.2 rev	Reverse sequencing primer from Thermo Scientific for sequencing of cloned inserts	AAGAACATCGATTTTCCATGGCAG
Up	Primer containing *tpiA* downstream homologous sequences as overhang for amplification of linear cassette	<u>GCAGAAGAAGTACCTGAGGCGCGGGAACATATGGGACGCTACGGGCTGGC</u>AATTAACCCTCACTAAAGGGC
Low	Primer containing *tpiA* upstream homologous sequences as overhang for amplification of linear cassette	<u>GTTGTCGCGGACAACGGCGAAAAGGGGCTGACCTTCGCTGTTGAACCGAT</u>TAATACGACTCACTATAGGGCTC
O (outside)	Primer for knockout verification binding 200 bp upstream of forward flanking homology region	TCGTCCTGATGTGCGCAAAC
23	Primer for knockout verification binding 200 bp downstream of reverse flanking homology region	AATGACCTGGCTACCCATCC
kil-left	Forward primer for amplification of *kil* target fragment for real-time PCR	CGCGATAAACCTCCTTGTTG
kil-right	Reverse primer for amplification of *kil* target fragment for real-time PCR	GATCCCCACCAATTCAGAAG
rpoS-left	Forward primer for amplification of *rpoS* target fragment for real-time PCR	GCACGTGAGTTGTCCCATAA
rpoS-right	Reverse primer for amplification of *rpoS* target fragment for real-time PCR	TAAGACGAAGCATACGGCTG
rpoD-left	Forward primer for amplification of *rpoD* target fragment for real-time PCR	TCGTATGTCCATCGGTGAAG
rpoD-right	Reverse primer for amplification of *rpoD* target fragment for real-time PCR	GCCACGGTTGGTGTATTTCT
bgl-left	Forward primer for amplification of *bgl* target fragment for real-time PCR	CTGACAAGTCCGTCTTATAAC
bgl-right	Reverse primer for amplification of *bgl* target fragment for real-time PCR	CAATCGAAAGCATAGGTATGG
leuB-left	Forward primer for amplification of *leuB* target fragment for real-time PCR	TGACCGGCGGCATCTATTTC
leuB-right	Reverse primer for amplification of *leuB* target fragment for real-time PCR	TCACTTTGTGGCGACGCTTG
tpiA-left	Forward primer for amplification of *tpiA* target fragment for real-time PCR	AATTCGCGGTGCTGAAAGAG
tpiA-right	Reverse primer for amplification of *tpiA* target fragment for real-time PCR	TCGATCTGACGTGCGCAAAC
tetR-left	Forward primer for amplification of *tetR* target fragment for real-time PCR	GCTGCTTCCTAATGCAGGAG
tetR-right	Reverse primer for amplification of *tetR* target fragment for real-time PCR	GGCACCTGTCCTACGAGTTG

gapA-left	Forward primer for amplification of *gapA* target fragment for real-time PCR	CGTTAAAGGCGCTAACTTCG
gapA-right	Reverse primer for amplification of *gapA* target fragment for real-time PCR	TAGCGTGAACGGTGGTCATC
bla-right	Forward primer for amplification of *bla* target fragment for measurement of relative plasmid abundance	CAACGATCAAGGCGAGTTAC
bla-left	Reverse primer for amplification of *bla* target fragment for measurement of relative plasmid abundance	TAACACTGCGGCCAACTTAC
41	Left traversing primer for inverse PCR to change promoter for *gfp* on pET-24a(+)-gfp+-lva	GCCGCTTATACTTGTGGCAAAT GAGCGGATAACAATTCCCCTC
42	Right traversing primer for inverse PCR to change promoter for *gfp* on pET-24a(+)-gfp+-lva	AAATCGGGTTACGCCGGAGAG CACGCCCTGCACCATTATGTTC

4.1.4 Storage of strains

For short-term storage, colonies or cultures were streaked onto LB-agar plates supplemented, when necessary, with the corresponding antibiotics. For long-term storage, bacteria were grown overnight in shake flasks in Terrific Broth (TB) medium. For auxotrophic strains, the culture was centrifuged in sterile polypropylene tubes (50 mL) and the supernatant removed. The cells were resuspended in autoclaved MilliQ water. Cells were then preserved by mixing 800 μL of bacterial culture with 200 μL of autoclaved 87% glycerol in sterile Eppendorf tubes. The tubes were then incubated at 4 °C for 10 min, followed by freezing with liquid nitrogen and storage at -80 °C.

4.2 Growth media

Complex medium was primarily used for molecular biological applications, whereas the supplemented glycerol ammonium sulphate (SGA) medium was used for fermentations and growing auxotrophic strains. The following sections give the compositions of the various media used in this work.

LB medium

Lysogeny broth (LB) was a nutrient medium developed originally for the study of bacteriophage infection of *Shigella dysenteriae* (Bertani, 1951). It allows rapid growth and freedom to cultivate in solid state with equal ease as in liquid medium and has become one of the most popularly used nutrient medium for bacterial cells. Table 4.4 gives the composition for LB medium used in this work.

Table 4.4: Composition of LB medium.

Component	Concentration (g L^{-1})
Soy peptone	10
Yeast extract	5
Sodium chloride	10

The components were dissolved in MilliQ water and the pH set to 7.4. LB medium for agar plates included Agar-Agar at a concentration of 15 g L^{-1} which was added after setting the pH. The medium was then sterilized by autoclaving.

TB medium

Terrific broth (TB) is a complex medium which was used for cultivation of cells to a higher biomass concentration than that achievable with LB. The composition for this medium is given under Table 4.5.

Table 4.5: Composition of TB medium.

Component	Ingredients	Amount (g)	Final concentration (g L^{-1})
Solution I (450 mL)	Casein pepton	6	12
	Yeast extract	12	24
Solution II (50 mL)	KH$_2$PO$_4$	1.2	2.4
	K$_2$HPO$_4$	6.25	12.5
Solution III	Glycerol (1000 g L^{-1})	2	3.9

The individual solutions were prepared and autoclaved separately and then mixed together under sterile conditions.

SGA medium

The supplemented glycerol ammonium sulphate (SGA) medium is a chemically defined medium described by Korz *et al.* (1995). The composition used in this work is based on the modified version by Sommer *et al.* (2009). The medium comprised of trace elements, salts solution, carbon source, nitrogen source, magnesium source and thiamine. Accordingly, the following stock solutions given under Table 4.6 were prepared.

Table 4.6: Stock solutions for the preparation of SGA medium.

Component	Ingredients	Concentration (g L^{-1})
Micronutrients solution 500x	$FeCl_3 \cdot 6H_2O$	5.4
	$ZnSO_4 \cdot 7H_2O$	1.38
	$MnSO_4 \cdot H_2O$	1.85
	$CoSO_4 \cdot 7H_2O$	0.56
	$CuCl_2$	0.17
	H_3BO_3	1
	$Na_2MoO_4 \cdot 2H_2O$	2.5
	Citric acid monohydrate	5
Salts solution 10x	K_2HPO_4	135.3
	KH_2PO_4	66.2
	Citric acid monohydrate	18.4
	EDTA	0.084
Carbon and energy source	Glycerol	1000
Nitrogen source	Ammonium sulphate	500
Magnesium source	$MgSO_4$	250
Thiamine	Thiamine	2

The stock solutions were prepared separately and sterilized by autoclaving with the exception of thiamine which was filter-sterilized with a Millex-GP 0.22 µm filter (Merck Millipore, Ireland). To MilliQ water of corresponding volume, SGA media components from the stock were mixed together according to the combination given under Table 4.7 in order to prepare the SGA medium.

Table 4.7: Preparation of SGA medium from the stock solutions of individual components

Component	Volume per L (mL)	Ingredients	Final concentration (g L^{-1})
Micronutrients solution 500x	2	$FeCl_3 \cdot 6H_2O$	0.0108
		$ZnSO_4 \cdot 7H_2O$	0.00276
		$MnSO_4 \cdot H_2O$	0.0037
		$CoSO_4 \cdot 7H_2O$	0.00112
		$CuCl_2$	0.00034
		H_3BO_3	0.002
		$Na_2MoO_4 \cdot 2H_2O$	0.005
		Citric acid monohydrate	0.01
Salts solution 10x	100	K_2HPO_4	13.53
		KH_2PO_4	6.62
		Citric acid monohydrate	1.84
		EDTA	0.0084
Carbon and energy source	10	Glycerol	10
Nitrogen source	12	Ammonium sulphate	6
Magnesium source	2.4	$MgSO_4$	0.6
Thiamine	5	Thiamine	0.01

Glycerol was present at a final concentration of 10 g L^{-1} and ammonium sulphate at 6 g L^{-1} for shake flasks and batch fermentations. For continuous cultivation, the glycerol concentration was 20 g L^{-1} in the inlet feed medium while all other components were maintained at the same concentration as in batch.

4.3 List of solutions

In the following tables, the compositions of the buffers and solutions used in this work are listed.

Table 4.8: Buffer 1 for preparation of chemical competent cells

Component	Concentration (mM)
CaCl$_2$	100
Tris-HCl	2

The pH was set to 7.4.

Table 4.9: Buffer 2 for preparation of chemical competent cells

Component	Concentration (mM)
CaCl$_2$	100
Tris-HCl	2
Glycerol	10 %

The pH was set to 7.4.

Table 4.10: Sodium acetate buffer (1 L)

Component	Concentration	Amount (g)
Sodium acetate	40 mM	3.28
Calcium chloride	20 mM	2.94
Bovine Serum Albumin (BSA)	50 µg mL^{-1}	0.05
Sodium azide	0.2 g L^{-1}	0.2

The pH of the buffer was set to 5.6.

Table 4.11: Binding buffer for IMAC (Buffer A)

Component	Concentration (mM)
Sodium phosphate	20
Sodium chloride	500
Imidazole	30

The pH was set to 7.4.

Table 4.12: Elution buffer for IMAC (Buffer B)

Component	Concentration (mM)
Sodium phosphate	20
Sodium chloride	500
Imidazole	500

The pH was set to 7.4.

Table 4.13: Nickel stripping buffer for IMAC

Component	Concentration (mM)
EDTA	50
Sodium phosphate	20
Sodium chloride	500

The pH was set to 7.4.

Table 4.14: TE Buffer

Component	Concentration (mM)
Tris-Cl	10
EDTA	1

Stock solutions of 1 M Tris and 0.5 M EDTA were set to pH 8.0 with HCl and NaOH respectively and used to prepare TE buffer.

Table 4.15: SRL solution for Eckhardt gel screening

Component	Concentration
Sucrose	20 %
RNase A	40 μg mL^{-1}
Lysozyme	2 mg mL^{-1}

Table 4.16: Tris-Glycine buffer

Component	Concentration
Tris	25 mM
Glycine	192 mM
SDS	0.1 %

Table 4.17: Laemmli sample loading buffer (2x)

Component	Concentration
Tris-HCl	250 mM
SDS	4 %
Glycerol	20 %
β-Mercaptoethanol	10 %
Bromophenol blue	0.002 %

Table 4.18: SDS-PAGE staining solution

Component	Concentration
Coomassie Brilliant Blue G-250	2 g L^{-1}
Acetic acid	10 % (v/v)
Isopropanol	25 % (v/v)
Water	65 % (v/v)

Table 4.19: SDS-PAGE destaining solution

Component	Concentration
Acetic acid	10 % (v/v)
Isopropanol	25 % (v/v)
Water	65 % (v/v)

Table 4.20: TE buffer-lysozyme mixture for RNA isolation

Component	Concentration
Tris-Cl	10 mM
EDTA	1 mM
Lysozyme	1 mg mL^{-1}

The pH of 1 M Tris-Cl and 0.5 M EDTA stock solutions were individually set to 8.0. DEPC-treated water was used.

Table 4.21: Tris-acetate EDTA buffer (50x)

Component	Amount
Tris	242 g
Acetic acid	57.1 mL
EDTA	100 mL

The final volume of the stock solution is 1 L. EDTA is a solution with a concentration of 0.5 M and a pH of 8.0.

Table 4.22: Tris-Borate EDTA buffer (5x)

Component	Concentration
Tris	445 mM
Boric acid	445 mM
EDTA	10 mM

The buffer was prepared in DEPC-treated water. The pH was adjusted to 8.3.

Dinitrosalicyclic acid reagent:

5 g of 3,5-Dinitrosalicylic acid was dissolved in 300 mL of distilled water at 50 °C by constant stirring. Upon reaching a homogeneous solution, 50 mL of 4 M sodium hydroxide was added and after continued stirring, 150 g of potassium sodium tartrate (Rochelle salt) was added to the solution. The

solution was stirred until the solids dissolved completely and allowed to cool. After cooling, the solution was transferred to a brown bottle to minimize light permeation and stored at room temperature.

Antibiotics

Ampicillin (sodium salt) and kanamycin sulphate (both from Carl Roth GmbH, Germany) were prepared as stock solutions of 50 mg mL^{-1} by dissolving in MilliQ water and filter-sterilized using a 0.22 μm filter (Millex-GP; Merck Millipore, Ireland). Working concentrations for ampicillin and kanamycin were 100 μg mL^{-1} and 50 μg mL^{-1} respectively. Tetracycline stock with a concentration of 5 mg mL^{-1} was prepared in 75% ethanol. Working concentration for tetracycline was 3 μg mL^{-1}.

L-Arabinose

Stock of 10% concentration in autoclaved MilliQ water was prepared and sterile-filtered into aliquots and stored at -20 °C.

4.4 Instruments, chemicals and consumables

Table 4.23: List of instruments used

Instrument	Manufacturer
Centrifuge microcentrifuge 5415R; Rotor F-45-24-11	Eppendorf AG, Germany
Centrifuge Sigma cooling centrifuge; 6K15; Rotor 12500	Sigma Laborzentrifugen GmbH, Germany
Centrifuge Sigma Microfuge 1-15; Rotor 12124	Sigma Laborzentrifugen GmbH, Germany
Chromatographic column XK26	Pharmacia Biotech, Sweden
Chromatography system (preparative) ÄKTAprime plus	GE Healthcare Europe GmbH, Germany
Deep freezer	Sanyo Electric Co. Ltd., Japan
Degasser for HPLC Erma ERC 3612	ERMA CR. Inc., Japan
DO electrode (P52201012)	Mettler-Toledo (Schweiz) GmbH, Switzerland
Dry constant temperature heater Dri-Block® DB-3D	Bibby Scientific Ltd., UK
Electroporator Gene Pulser® II	Bio-Rad Laboratories GmbH, Germany
Gel documentation system BioDocAnalyze	Biometra GmbH, Germany
HPLC autosampler Marathon	Spark Holland B.V., The Netherlands
HPLC column Nucleogel sugar 810H; cation exchange	Macherey-Nagel GmbH, Germany
HPLC column thermostat ERC Gecko-2000	ERC GmbH, Germany
HPLC guard column Nr. 719537	Macherey-Nagel GmbH, Germany
HPLC pump Irica Pump Σ871	Irica Instruments Inc., Japan
Ion-exchange column for preparation of MilliQ water Seralpur Pro 90 CN	SERAL Reinstwassersysteme GmbH, Germany

Ion meter portable P507	Consort bvba, Belgium
NanoDrop® 1000 spectrophotometer	Peqlab Biotechnologie GmbH, Germany
Orbital shaker Lab-Shaker LS-X	Adolf Kühner AG, Switzerland
pH electrode (405-DPAS-SC-K8S)	Mettler-Toledo (Schweiz) GmbH, Switzerland
pH meter 691	Metrohm, Switzerland
Power source for protein electrophoresis Power Pack P 20	Biometra GmbH, Germany
Pump for chemostat inlet feed Ismatec MV-CA/04	IDEX Health and Science GmbH, Switzerland
Pump for chemostat outlet suction MBR Bioreactor AG PP101-2	Sulzer Pumps, Switzerland
Pump for feeding in fed-batch	Watson-Marlow; Bioengineering AG, Switzerland
Real-time PCR cycler Rotor Gene RG-3000	Corbett Research, Australia
Sonicator Branson Sonifier 450	Branson Ultrasonics, USA
Spectrofluorophotometer RF-5301PC	Shimadzu Corporation, Japan
Thermal cycler for PCR Mastercycler® proS	Eppendorf AG, Germany
Thermal cycler for PCR Bioer LifeTouch; TC-96/G/H(b)B	Hangzhou Bioer Technology Co. Ltd., PRC
UV-visible spectrophotometer Biochrom 4060	Pharmacia Biotech, Sweden
Vacuum drying oven Heraeus VT 5042 EK	Heraeus, Germany
Vortexer Vortex Genie 2	Scientific Industries Inc., USA
Waterbath Julabo F10	JULABO GmbH, Germany
Water bath for boiling IKA Ter 2	IKA Labortechnik, Germany
Weighing balance Mettler AE260	Mettler Instrumente GmbH, Switzerland

Table 4.24: List of chemicals used

Chemical	Supplier
Acetic acid	VWR Int., France
Acrylamide (0.8 % N,N´-bis-methylene acrylamide / 30 % acrylamide mixture)	Carl Roth GmbH, Germany
Agar-Agar	Carl Roth GmbH, Germany
Agarose	Carl Roth GmbH, Germany
Ammonium persulfate	Carl Roth GmbH, Germany
Ammonium sulphate	Carl Roth GmbH, Germany
Ampicillin (sodium salt)	Carl Roth GmbH, Germany
Arabinose (-L)	Carl Roth GmbH, Germany
Boric acid	VWR Int., Belgium
Calcium chloride dihydrate	Applichem GmbH, Germany
Casein peptone	Carl Roth GmbH, Germany
Cobalt sulphate heptahydrate	Merck KgaA, Germany
Coomassie Blue G 250	Carl Roth GmbH, Germany
DEPC	Sigma-Aldrich Chemie GmbH, Germany
Dipotassium hydrogen phosphate	Applichem GmbH, Germany
dNTP mixture	Thermo Scientific, Germany

DNSA (3,5-Dinitrosalicylic acid)	Sigma-Aldrich Chemie GmbH, Switzerland
Ethanol, absolute	VWR Int., France
Ethylenediaminetetraacetic acid, disodium salt, dihydrate	Carl Roth GmbH, Germany
Ferric chloride	Merck KgaA, Germany
Glycerol	Emery Oleochemicals GmbH, Germany
Imidazole	Applichem GmbH, Germany
Isopropanol	VWR Int., France
Kanamycin sulphate	Carl Roth GmbH, Germany
Magnesium sulphate	Carl Roth GmbH, Germany
Manganese sulphate monohydrate	Merck KgaA, Germany
Pluronic PE8100 (Antifoam)	BASF, Germany
Potassium dihydrogen phosphate	Th.Geyer GmbH, Germany
Potassium sodium tartrate tetrahydrate	Sigma-Aldrich Chemie GmbH, Germany
Sodium dodecyl sulphate	Carl Roth GmbH, Germany
Sodium hydroxide	VWR Int., France
Sodium molybdate dehydrate	J.T Baker, Holland
Soy peptone	UD Chemie GmbH, Germany
TEMED (N-N-N´-N´-tetramethylethylenediamine)	Carl Roth GmbH, Germany
Tetracycline	Fluka Chemie GmbH, Switzerland
Thiamine hydrochloride	Sigma-Aldrich Chemie GmbH, Germany
Tris	Carl Roth GmbH, Germany
Yeast extract	Ohly GmbH, Germany
Sucrose	Applichem GmbH, Germany
Sulphuric acid	VWR Int., France
Urea	Carl Roth GmbH, Germany

Table 4.25: List of consumables

Product	Supplier
Disposable cuvettes, 1.5 mL, semi-micro	Brand GmbH, Germany
Electroporation cuvette, type LE (Long Electrode; 1 mm gap)	Peqlab Biotechnologie GmbH, Germany
Filter for sterilization Millex-GP, 0.22 μm	Merck Millipore, Ireland
Glass vials for HPLC, 1.5 mL	Techlab GmbH, Germany
Microtubes, 1.5 mL, 2 mL	Eppendorf AG, Germany
Pipette tips	Greiner Bio-one GmbH, Germany
Polypropylene tubes, sterile, 15 mL or 50 mL	Greiner Bio-One GmbH, Germany
RNase-free filter-tips 1 mL, 200 μL, 20 μL	STARLAB GmbH, Germany
RNase-free microfuge tubes 1.5 mL/ 2 mL	STARLAB GmbH, Germany
RNase-free PCR tubes 0.2 mL	STARLAB GmbH, Germany

RNase-free PCR tubes 0.1 mL	Biozym Scientific GmbH, Germany
Silicon tubes for bioreactor 3 mm, 6 mm, 10 mm i.d.	Kahmann & Ellerbrock GmbH & Co.KG, Germany
Silicon tubes for chemostat feed 1mm i.d.	VWR, Germany

4.5. Molecular genetic methods

4.5.1 Plasmid DNA isolation

Plasmid DNA was isolated using the kit Wizard® *Plus* SV Minipreps DNA Purification System (Promega GmbH, Germany) according to the instructions provided by the manufacturer. The isolation based on the principle of alkaline lysis is combined with the efficiency of silica membranes for the purification of plasmid DNA. Overnight grown bacterial culture samples were centrifuged in a Sigma 1-15 microfuge and resuspended in a solution containing 50 mM Tris-HCl (pH 7.5) and 10 mM EDTA. Additionally, the presence of RNase A ensured removal of contaminating RNA. The cells were lysed under highly alkaline conditions using the lysis solution containing 0.2 M NaOH and 1% SDS. Both plasmid and genomic DNA were denatured, but only plasmid DNA was able to renature correctly upon addition of the neutralization solution containing 0.759 M potassium acetate and 2.12 M glacial acetic acid. The lysate was separated and added onto a silica-based spin column wherein the binding of DNA was improved due to the presence of guanidine hydrochloride in the neutralization solution. The bound DNA was washed with a solution containing 60% ethanol and eluted as pure plasmid with autoclaved MilliQ water.

The concentration of the isolated DNA was estimated based on its absorption at 260 nm using a NanoDrop® 1000 spectrophotometer (Peqlab Biotechnologie GmbH, Germany). The purity of the DNA was verified by measuring the ratio of the absorbances at 260 nm and 280 nm which related to protein contamination and the ratio at 260 nm and 230 nm that showed up contaminations by other compounds such as EDTA, guanidium salts, phenol and agarose.

DNA preparations were analysed by electrophoresis on a 1% agarose gel in 0.5x working concentration of tris-acetate-EDTA buffer (see chapter 4.3). Visualization of the DNA was possible by addition of the dye Roti-Safe® GelStain (Carl Roth GmbH + Co. KG, Germany) at a ratio of 5 µL per 100 mL agarose gel and observing the fluorescence of the DNA bands using the BioDocAnalyze gel documentation system (Biometra GmbH, Germany).

4.5.2 Genomic DNA isolation

Genomic DNA from bacterial cells was isolated using the Wizard® Genomic DNA Purification Kit (Promega GmbH, Germany) by following the protocol provided by the manufacturer for Gram negative bacteria. The procedure involved the lysis of pelleted bacterial cells using the Nuclei Lysis solution (proprietary composition containing Tris base and SDS) and incubation at 80 °C for 5 min. Thereafter, the sample was cooled down to room temperature and incubated with RNase A at 37 °C for

1 h to remove any contaminating RNA. The proteins in the sample were then salt-precipitated and the clear lysate with solubilized genomic DNA was precipitated with isopropanol and further purified by washing with 70 % ethanol. The dried DNA pellet was solubilized in a Tris-EDTA buffer that had been warmed up to 50 °C.

4.5.3 Polymerase Chain Reaction

The Phusion High-Fidelity DNA polymerase kit (Fisher Scientific - Germany GmbH) was used for polymerase chain reaction. This enzyme represented an important step ahead of the conventionally used Taq DNA polymerase which had a high frequency of error (about 1 for every 9000 nucleotides). Phusion contains 5' → 3' polymerase activity as well as 3' → 5' exonuclease activity and generates blunt end PCR products. With genomic DNA templates, Phusion polymerase typically needed 30 s per kb for elongation, whereas with plasmid DNA templates, the speed goes up to 15 s per kb. The Phusion HF buffer was predominantly used for all amplifications whereas the alternative GC buffer was used in case of long amplicons or when the optimal amplification conditions had to be found and optimized. Primers in lyophilized form and of 'desalted' purity grade (Invitrogen, Germany) were reconstituted to 100 µM stock concentration with nuclease-free water (Promega GmbH, Germany). Thermal cycling was carried out using the Mastercycler proS instrument (Eppendorf AG, Germany) or the Bioer LifeTouch PCR machine (Hangzhou Bioer Technology Co. Ltd., PRC). After the cycles were completed, PCR reactions were checked for successful amplification by agarose gel electrophoresis. Preparative reactions were pooled together after verification of a small aliquot by electrophoresis. Amplified DNA from pooled samples were purified using the Wizard® SV Gel and PCR Clean-Up System (Promega GmbH, Germany).

4.5.4 Inverse PCR

The inverse PCR method (Imai et al., 1991) was used to modify sequences in existing plasmids for example to change the promoter upstream of a coding sequence in order to alter the transcriptional activity of the gene. Primers were designed in such a way that they oriented in opposite directions on the template plasmid and amplified the complete molecule while ignoring the targeted promoter. Each of the primers carried one half of the new promoter sequence to be added so that the amplified linear molecule would have the two halves of the new promoter at its ends (shown later in Fig. 5.41). Since the primers were not phosphorylated at their 5' ends, the linear amplicon was purified, phosphorylated using T4 polynucleotide kinase and self-ligated using T4 DNA ligase (both Thermo Scientific, Germany). The recircularized plasmid was transformed into competent cells of E. coli DH5α or E. coli Top10.

4.5.5 Restriction digestion

DNA molecules were restricted using restriction endonucleases (Thermo Scientific, Germany) for various objectives like preparation of vectors and inserts in cloning, screening for positive clones, testing the presence or absence of sites or degradation of template plasmid after inverse PCR.

Generally, the corresponding reaction buffer was used with an incubation time of 3 h at 37 °C. Any deviation from this general scheme was designed as necessary for the particular enzyme in question and according to guidelines from the manufacturer. For FastDigest® enzymes, an incubation time of 15 min was sufficient. For plasmid digestions, 0.2 μg to 1 μg of DNA was added to a reaction volume of 20 μL. When preparing vectors for cloning, the restricted plasmid was dephosphorylated at the 5' ends by using calf intestine alkaline phosphatase (Fermentas, Lithuania) or FastAP (Thermo Scientific, Germany). Restricted DNA was electrophoresed on agarose gel and documented for analysis or eluted from the gel and further purified for preparative applications.

4.5.6 DNA elution

After preparative restriction and gel electrophoresis, DNA was observed under reduced UV intensity (50 %) in the BioDocAnalyze gel documentation system and quickly cut out using a fine scalpel. A reduced agarose concentration of 0.8 % was used for these applications in order to minimize impurities in the final DNA preparation. The DNA from the excised agarose gel was eluted using the kit Wizard® SV Gel and PCR Clean-Up System (Promega GmbH, Germany). The agarose slice containing the DNA of interest was dissolved at a temperature of 50 °C in a buffer containing guanidine isothiocyanate. In the presence of this chaotropic salt, the binding of the released DNA onto a silica membrane in a spin column was promoted. After washing the bound DNA with a buffer containing 80% ethanol, pure DNA was eluted with nuclease-free water and checked for its yield and purity using the NanoDrop® 1000 spectrophotometer (Peqlab Biotechnologie GmbH, Germany).

4.5.7 Ligation

T4 DNA ligase (Thermo Scientific, Germany) was used for cohesive- or blunt-end ligation including self-ligation for recircularization of linear amplicons according to the guidelines from the manufacturer. For cohesive-end ligations, an insert to vector molar ratio of 3:1 was used. Optimization of blunt-end ligation was possible by addition of PEG 4000 (Thermo Scientific, Germany) to the reaction. For blunting cohesive ends prior to ligation, Klenow fragment (Thermo Scientific, Germany) was used. The reactions were incubated for 3 h or overnight at 4 °C and subsequently used for transformation. Chemical competent cells (50 μL) were transformed with 5 μL or 10 μL of the ligation mixture whereas in the case of electroporation, the ligase was first inactivated by incubation at 70 °C for 5 min and subsequently 3 μL of the mixture was used for electroporation of 50 μL of electrocompetent cells.

4.5.8 Competent cells preparation and transformation

Chemical-competent cells of *E. coli* were prepared by the calcium chloride method. An overnight culture in LB medium at 37 °C was first established. This was used to inoculate the main culture which consisted of a 1 L shake-flask containing 250 mL LB medium. Upon reaching an optical density (OD_{600}) of 0.9, the cultivation was stopped in order to procure cells in their exponential growth phase. The cells were incubated at 0 °C for 15 min followed by centrifugation for 10 min at 4000g and 4 °C temperature in a cooling centrifuge (Sigma 6K15; Sigma Laborzentrifugen GmbH, Germany). While keeping the cells cooled throughout the procedure, the supernatant was removed and the cell pellet gently resuspended in 10 mL of ice-cold Buffer 1 (see chapter 4.3). The cells were centrifuged again under the same conditions as mentioned earlier and the supernatant removed. The pellet was resuspended in 10 mL of ice-cold Buffer 2 (see chapter 4.3) and aliquoted as 50 µL portions into sterile 1.5 mL tubes (Eppendorf AG, Germany). The aliquots were then frozen with liquid nitrogen and stored at -80 °C until further use.

Chemical competent cells were transformed by the heat-shock method. Frozen competent cells were thawed on ice. For plasmid transformation, 10-50 ng of DNA was added to the cells, briefly mixed, and the cells quickly returned to ice where they were incubated for 20 min. After incubation, the cells were subjected to a heat shock at 42 °C in a waterbath for exactly 45 s and then returned immediately to ice and incubated for 5 min. Subsequently, 1 mL of LB medium at room temperature was added to the cell suspension and the tubes were incubated at 37 °C with shaking for 1 h in order to revive the cells. Finally, the cells were pelleted by centrifuging for 1 min at 2000g in a Sigma 1-15 microfuge (Sigma Laborzentrifugen GmbH, Germany) at room temperature. The supernatant was removed up to a residual volume of about 100 µL in which the cell pellet was resuspended, and the concentrated cell suspension was used for plating on LB-agar plates with the corresponding antibiotic for selection.

For preparation of electrocompetent cells, an overnight bacterial culture in LB was used to inoculate the main culture which consisted of two 1 L shake-flasks of 250 mL LB medium each. The cells were allowed to grow up to an optical density of 0.7 whereupon the growth was arrested by incubation on ice for 15 min. The cells were centrifuged for 15 min at 4000g and 0 °C in a cooling centrifuge (Sigma 6K15; Sigma Laborzentrifugen GmbH, Germany). The supernatant was removed carefully and any residual medium was washed out with 25 mL of ice-cold autoclaved MilliQ water without disturbing the cell pellet. The pellet was then resuspended in 5 mL of ice-cold autoclaved MilliQ water before increasing the volume up to 150 mL with ice-cold water. Both the suspensions were now mixed and centrifuged again under the same conditions as earlier. The supernatant was removed and the cells were resuspended in 10 mL of an ice-cold solution of 15% glycerol before increasing the volume to 40 mL with the glycerol solution. The cells were then centrifuged for 10 min at 4000g and 0 °C and the supernatant removed. The pellet was resuspended in 1 mL of an ice-cold solution of 10% glycerol by gentle shaking. The cells were finally aliquoted into 100 µL portions and frozen under liquid nitrogen for storage at -80 °C until further use.

For transformation of electrocompetent cells (electroporation), the Gene Pulser® II (Bio-Rad Laboratories GmbH, Germany) was used. Electrocompetent cells of 100 µL volume were thawed on ice. After addition of plasmid DNA or appropriate volume of a ligation reaction, the cells were incubated further on ice for 1 min and transferred to a pre-cooled electroporation cuvette of type LE (Long Electrode; 1 mm gap; Peqlab Biotechnologie GmbH, Germany). The cuvette was wiped dry and placed in the cuvette slide of the electroporator and the slide pushed into the shocking chamber until a firm contact was established between the electrodes of the chamber. Electroporation was carried out with the following settings: capacitance of 25 µF, resistance of 400 Ω and a voltage of 1.8 kV. Immediately after the voltage pulse, 1 mL of pre-warmed LB medium was added to the cells and the suspension incubated at 37 °C under shaking for 1 h in an orbital shaker (Adolf Kühner AG, Switzerland). After incubation, the cells were concentrated to a volume of 100 µL and plated onto LB-agar plates containing appropriate antibiotic for selection.

When ligation reactions would be used for electroporation, the presence of salt due to the ligation buffer was a critical issue. Therefore, the DNA was precipitated and concentrated from the 20 µL ligation mixture down to 2-3 µL. Firstly, 3 M sodium acetate (pH 5.2) was added to the ligation mixture to a final concentration of 0.3 M. Three volumes of chilled 100% ethanol were then added to the solution and mixed well. The solution was incubated at -20 °C for 30 min prior to the centrifugation of the DNA at 14,000g and 0 °C for 10 min. The resulting pellet was washed once with 70% ethanol and centrifuged again. The precipitated DNA pellet was dried until traces of ethanol could not be detected by smell. The DNA pellet was dissolved in 3 µL of autoclaved MilliQ water and used for transformation or stored at -20 °C.

4.5.9 Eckhardt gel screening

A rapid method for screening of large number of transformants was applied (Eckhardt, 1978). Colonies were picked up and resuspended in 10 µL TE Buffer and briefly mixed with 20 µL of a solution of sucrose, RNase and lysozyme (see chapter 4.3). The cell mixture was loaded onto an Eckhardt gel which was prepared by adding 1 mL of a 20% SDS solution to a 100 mL solution of agarose in TAE buffer just before pouring onto the cast. Cell lysis was allowed to take place in the wells of the gel for about 30 min after which the electrophoresis was started. Sizes of plasmid DNA released were compared with a control DNA sample containing the vector without insert.

4.5.10 *In silico* analysis and file management

All *in silico* work such as management of DNA sequences, restriction site analysis, primer design, creation of plasmid maps and sequence alignment were carried out using the program Clone Manager Professional Suite version 8 (Scientific and Industrial Software, USA). Primers for cloning were designed with the following basic criteria, allowing deviations only when absolutely necessary: GC content was always in the range of 50 % - 60 %. Melting temperature (T_m) was in the range of 55 °C – 80 °C for PCR primers and 30 °C – 100 °C for sequencing primers. Interactions between the primers at the 3' end was kept to less than 3 base pair bondings while interactions throughout the length of the molecule was kept to less than 7 base pair bondings. The last base was always either a G or a C in order to form the stable 'GC clamp' at the 3' end, and runs of any single base was limited to a maximum of 3 (for *e.g.*, AAA).

4.5.11 Genome engineering by Red/ET recombination

Quick and Easy *E. coli* Gene Deletion Kit (Gene Bridges GmbH, Germany) was used for deletion of a target gene on the *E. coli* genome to create knockout strains. The kit is based on the concept of Red/ET recombination which allows a precise and specific homologous recombination with a functional linear DNA cassette. This cassette contained an antibiotic resistance gene for selection of positive recombinants and flanking regions homologous to the upstream and downstream regions of the target gene. The kit contained 2 types of Red/ET expression plasmids which provided the phage proteins necessary for the recombination. Since the linear cassette was flanked by FRT sites, the resistance cassette from positive recombinants could be removed if required, through the application of a FLP-recombinase expression plasmid ultimately leaving behind only a scar sequence. These steps are shown schematically in Fig. 4.1.

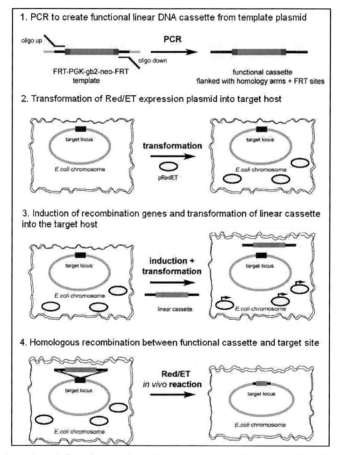

Fig. 4.1: Experimental work-flow for creation of a target-gene knockout in an *E. coli* genome through homologous recombination. Illustration adapted from technical protocols, Gene Bridges GmbH.

The plasmid pRedET (Appendix, Fig. 8.3) contained the genes coding for the enzymes involved in recombination (alpha, beta and gamma) under the control of the pBAD promoter which was induced by L-arabinose. The plasmid contained a pSC101 origin, the gene *repA* coding for a temperature-sensitive protein required for replication and antibiotic resistance against tetracycline or ampicillin depending on the plasmid type (see Table 4.2). At 30 °C this plasmid is maintained at 5 copies per cell while being unable to replicate at 37 °C.

The template plasmid pFRT (Table 4.2) containing the FRT-pgk-gb2-neo-FRT cassette (Fig. 8.4) is a 3446 bp circular plasmid with a R6K conditional origin of replication that allows it to replicate only in *E. coli* strains with the *pir* genotype. It contained no resistance gene other than the kanamycin resistance in the selection cassette (Michaela Biener, Gene Bridges GmbH, personal communication). The kanamycin resistance gene on the functional cassette is driven by the gb2 promoter for use in prokaryotes. An additional pgk promoter offered the possibility for expression in mammalian cells. A synthetic polyadenylation signal terminates the kanamycin resistance gene. Due to its conditional origin of replication, pFRT is not expected to replicate in the commonly used laboratory strains of *E. coli*. This origin requires the trans acting Π protein provided by the *pir*⁺ or *pir116* marker on the genome of the supporting strain in order to replicate. The marker *pir116* is characterized by a single base mutation causing 1 amino acid residue substitution in the Π protein, resulting in a higher copy number of the plasmid. Therefore, these plasmids are suicide vectors in *pir⁻* strains (Soubrier *et al.*, 1999).

Plasmid pFRT containing the kanamycin resistance gene was used as the template in a PCR to amplify this region using primers containing overhangs that were homologous to the flanking regions of the target site on the genome. The target *E. coli* strain was transformed with the pRed/ET expression plasmid and maintained at 30 °C. For recombination, the strain containing the pRed/ET expression plasmid was induced with 50 µL of L-arabinose stock solution (see chapter 4.3) per 1.4 mL of culture and incubated at 37 °C to express the Redα/Redβ proteins and also inhibit plasmid replication. Following the incubation, the functional linear cassette previously generated was electroporated into the target strain and recombination allowed to take place during incubation at 37 °C for 3 h with shaking. Following recombination, the cells were plated onto LB-agar plates containing kanamycin in order to select for positive recombinants. Potential recombinants were screened for using a combination of primers to distinguish between clones that contained a precise gene deletion, those with the target gene intact but also the resistance gene added on to the genome and those with no change in the genome that have been able to survive the selective conditions.

4.5.12 Cycle-sequencing

Potential clones were verified by sequencing the cloned region using corresponding sequencing primers. Cycle-sequencing was performed at the Sequencing Core Facility (SCF), Centre for Biotechnology (CeBiTec) at Bielefeld University, Germany. A BigDye® terminator version 3.1 chemistry was used for the PCR in a GeneAmp® PCR System 9700 (both Applied Biosystems, Germany). The fragments of various lengths were subsequently separated by capillary electrophoresis and the fluorescence signal of each base was converted to their corresponding digital data by a 96-well 3730*xl* DNA Analyzer (Applied Biosystems, Germany).

4.6 Microbial cultivation

4.6.1 Shake-flask cultivation

Bacteria were cultivated in a volume of 50 mL in shake flasks of 300 mL total volume (Duran Group GmbH, Germany). The flasks were equipped with autoclavable silicon caps (Omnilab-Laborzentrum GmbH & Co. KG, Germany) that allowed the sterile exchange of air. The cultivations were carried out at 37 °C or 30 °C depending upon the strain. Aeration and agitation were realized by a Lab-Shaker LS-X orbital shaker (Adolf Kühner AG, Switzerland) operated at a speed of 130 rpm and an eccentricity of 50 mm. All cultivations of the reference strain *E. coli* JM109 in SGA included thiamine (0.01 g L^{-1}).

4.6.2 Batch and Fed-batch cultivation

Batch and fed-batch experiments were conducted in a 7 L NLF 22 fermenter (Bioengineering AG, Switzerland) with 5 L working volume. Water and glycerol for the SGA medium (Table 4.7) were sterilized *in situ* along with the fermenter, while the remaining components were sterilized separately and added under sterile conditions. After adjusting initial pH and saturating the medium with oxygen, the fermenter was inoculated from an overnight shake flask culture. The air flow rate used was 1 vvm and the temperature was maintained at 37 °C for the reference strain *E. coli* JM109-p582 and at 30 °C for auxotrophic strains. Orthophosphoric acid (10%) and sodium hydroxide (2 M) were used for pH maintenance. CO_2 in the exhaust air was analysed using a GMP 221 CO_2 sensor (Vaisala Oyj, Finland). The variables measured at the Intelligent Front Module (IFM) in the fermenter were converted into digital signals for monitoring and control by the Profibus Handler program, and the complete bioprocess was run using the software BiOSCADA Lab (both from Bioengineering AG, Switzerland).

During batch experiments, the impeller speed was initially set to 200 rpm and later cascaded to the dissolved oxygen level (*DO*) when the latter fell below 60% saturation. In fed-batch mode, the impeller was initially operated at a basic speed of 200 rpm and the *DO* threshold set at 30% minimum. A concentrated feed solution of 700 mL volume was prepared for fed-batch which was a mixture of glycerol and ammonium sulphate both autoclaved separately and mixed together under sterile conditions. Final concentrations in the feed solution were 200 g L^{-1} glycerol and 60 g L^{-1} ammonium sulphate. During feeding, the impeller speed was maintained at the level finally reached, while the feed pump (Watson-Marlow) was cascaded to the *DO* with a maximum limit of 60% dissolved oxygen saturation.

Batch cultivations were also carried out in a 2 L in-house mini-fermenter (exact total volume 1.96 L) with a working volume of 1 L shown in the schematic diagram in Fig. 4.2. The reactor had a height to diameter ratio of about 3 and comprised of a stainless steel vessel bottom containing the water-jacket for temperature maintenance and a cylindrical glass wall placed above. Agitation was carried out at a stirrer speed of 800 rpm and aeration with a space velocity of 1 vvm. pH and *DO* were measured in-line by the use of a pre-pressurized autoclavable pH electrode and autoclavable *DO* electrode respectively (both from Mettler-Toledo (Schweiz) GmbH, Switzerland). The process was monitored by an ADAM-4060 relay output module (Advantech Ltd., USA) and run using the software DASYLAB 6.0 (National Instruments Service GmbH, Germany). The key specifications in the construction of this fermenter are listed under Table 4.26. Fig. 4.3 shows the mini-fermenter in operation.

Table 4.26: Technical specifications for the in-house mini-fermenter.

Definition of parameter	Size (mm)
Reactor diameter (D)	94.4
Reactor height (H)	280
Impeller diameter (d)	46
Height of impeller blades (a)	12
Width of impeller blades (b)	12
Diameter of disc (c)	30
No. of blades per impeller	6
No. of impellers	3
Distance between impellers (k)	45
Distance between impeller blade and sparger (m)	5 (approx.)
Distance between lowest impeller and reactor floor	45
Stirrer shaft diameter	8
No. of baffles	4
Height of baffle (g)	260
Width of baffle (e)	8
Separation of baffle from reactor wall (f)	2
Distance between baffle and reactor floor (h)	10

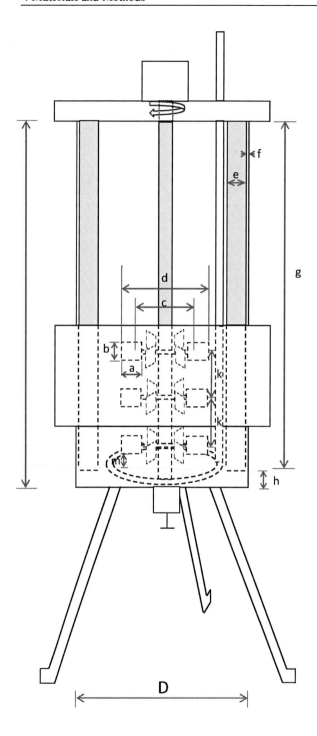

Fig. 4.2: Schematic representation of the 2 L in-house fermenter

Fig. 4.3: The 2 L mini-fermenter in operation.

4.6.3 Continuous culture and chemostat operation

Continuous culture experiments were carried out in the 2 L in-house fermenter with 1 L working volume. The process was started as a typical batch fermentation and allowed to run until the end of the exponential growth phase. The feed and outlet pumps (see chapter 4.4) were then switched on and the desired space velocity was achieved by adjusting both the pumps so that the culture level in the reactor always remained constant. The feed medium supplied from a 40 L reservoir had the same composition as the SGA medium for batch but with 20 g L^{-1} glycerol (see Table 4.7).

To avoid any errors in fluid suction through the tubes that could accumulate over long periods of time and result in gross volume variations, the outlet pump was always operated at 2-3 times the rate of the inlet pump and the suction pipe entrance was positioned at a fixed depth corresponding to a liquid working volume of 1 L. A steady state was attained by running the process at properly maintained constant environmental conditions like pH, temperature, aeration rate, agitation, culture volume, and inlet feed rate. Stably maintained levels of dissolved oxygen (*DO*) in the culture medium, CO_2 in the exhaust air stream and optical density of the culture were assumed to indicate a steady state. Any change in the feed flow rate was followed by allowing the elapse of 3-4 residence times before a new steady state was assumed. With the control strain *E. coli* JM109-p582, when antibiotic was applied for selection pressure, it was either administered as a pulse during the process or already added to the feed medium.

4.7 Bioprocess analytics

4.7.1 Substrate carbon and nitrogen source measurement

Culture samples were centrifuged to separate the cells. Residual glycerol concentration (carbon and energy source) in the supernatant was measured by HPLC using a Nucleogel sugar 810H cation exchange column (Macherey-Nagel GmbH, Germany). The column had a length of 300 mm and an internal diameter of 7.8 mm. The stationary phase was made up of a sulfonated, spherical polystyrene/divinylbenzene polymer matrix in H^+ form with the particle diameter ranging between 7 μm and 11 μm. Sulphuric acid at a concentration of 2.5 mM was used with a flow rate of 0.8 mL min^{-1} as the mobile phase. The analytes were detected by an ERC- 7515A refractive index detector (ERMA CR. Inc., Japan).

The temperature of the column was maintained at 70 °C using an ERC Gecko-2000 thermostat (ERC GmbH, Germany). Standard glycerol solutions of 5 g L^{-1}, 10 g L^{-1} and 20 g L^{-1} were analyzed in duplicate (Appendix, Table 8.3). For the analysis, 500 μL of standard solution or centrifuged culture supernatant was filled in 1.5 mL glass vials (Techlab GmbH, Germany) from which 20 μL was drawn through the column by a Marathon autosampler (Spark Holland B.V., The Netherlands). The elution of the analytes from the column was isocratic. The pump Irica Σ871 (Irica Instruments Inc., Japan) delivered a pressure of 4-5 MPa during operation. The process was controlled using the software ChromStar Light version 6.3 (SCPA GmbH, Germany).

Residual ammonium sulphate in the extracellular medium was estimated through detection of the corresponding amount of ammonia released after addition of 10 M NaOH, using a NH500/2 ion-selective electrode (WTW GmbH, Germany) and measuring the negative voltage using a P507 portable ion meter (Consort bvba, Belgium). A calibration curve was obtained from standard ammonium solutions of 0.01 g L^{-1}, 0.1 g L^{-1} and 1 g L^{-1} prepared with the corresponding amounts of ammonium sulphate (Appendix, Table 8.1). For the calibration, 1 mL of standard solution was added to 10 mL of distilled water and stirred continuously. The liberation of ammonia was initiated by the addition of 150 μL of 10 M NaOH and the voltage reading was noted after it reached a stable value. The samples were diluted at a ratio of 1:10 in water before analysis by the same procedure as for the standard. The ammonium ion concentration in the sample was read from the standard curve and the corresponding ammonium sulphate concentration could then be calculated.

4.7.2 Biomass quantification

The growth of bacterial cell population was monitored by analyzing the blocking of the incident light by the cell suspension in a sample which was taken as its optical density (OD_{600}) at 600 nm measured using a Biochrom 4060 UV-visible spectrophotometer (Pharmacia Biotech, Sweden). Culture samples were diluted in water so that the measured OD_{600} did not exceed 1.0 since the linear correlation gets lost at higher densities. An optical density of 1 OD_{600} was found to correspond to a bacterial cell number concentration of 10^9 mL^{-1}.

To find the dry biomass concentration, 2 mL of bacterial culture was taken in an Eppendorf tube whose mass had been weighed previously. The sample was centrifuged at $12000g$ for 10 min in a Sigma 1-15 microfuge (Sigma Laborzentrifugen GmbH, Germany) and the supernatant removed. The pellet was washed with 1 mL of saline solution (0.9% NaCl) and centrifuged again. The wet cell pellet was then dried overnight in a Heraeus vacuum drying oven at 60 °C and a pressure of < 50 mbar (Heraeus, Germany). The dried cell pellet was weighed and the difference in the masses was used to calculate the dry biomass concentration. Results were represented graphically using the software OriginPro version 8.5.0 SR1 (OriginLab Corporation, USA) or Microsoft Excel 2010 (Microsoft Corporation, USA).

4.7.3 Enzyme activity assay for β-glucanase

The test for measuring the activity of endo-1,3-1,4-β-glucan-glucanohydrolases was based on the hydrolysis of lichenan (Sigma-Aldrich Chemie GmbH, Germany) which is a β-glucan polymer containing β-1,4- and β-1,3-bonds in a proportion of 70% and 30% respectively. Apart from lichenan, barley β-glucan (Megazyme International, Ireland) which has a very similar composition of β-linkages was also used. The maximum activities observed with an excess of pure enzyme were found to be comparable with both the substrates. The original study detailing the construction of the hybrid β-glucanase gene H1 also describes the enzyme activity on barley β-glucan as well as lichenan as substrates (Borriss et al., 1989). 1 Unit is defined as the amount of enzyme activity that releases 1 μmol of reducing ends per min at 50 °C temperature and pH 5.6.

For the assay, 50 mg of the substrate was dissolved in 10 mL of sodium acetate buffer (Table 4.10) and the solution boiled for 10 min. To determine the enzyme activity in the extracellular medium, cell-free supernatants were prepared by centrifugation of the culture samples. The supernatants were diluted with the sodium acetate buffer on the basis of their cell concentrations in order to remain within the linear range of the assay. If the substrate solution had been prepared earlier and stored at 4 °C, it was warmed for 10 min at 50 °C and then 200 μL per reaction was added to 1.5 mL Eppendorf tubes. To this solution, 20 μL of the diluted supernatant was added, mixed briefly by vortexing and incubated for 20 min at 50 °C. For the blank reaction, 20 μL of buffer was added in place of the sample, whereas for the test reaction, 20 μL of an IMAC-purified fraction of the enzyme (see Chapter 4.7.7) was added. The test reaction measured the maximum concentration of reducing end groups achieved with the particular substrate preparation at maximum conversion. At the end of the incubation, the reactions were stopped by briefly transferring the tubes on ice prior to analyzing the reducing end groups released by addition of 100 μL of dinitrosalicylic acid reagent (see chapter 4.3). The tubes were vortexed, spun down briefly and perforated with a needle. The tubes were transferred to a boiling water bath for 10 min. At the end of boiling, 1 mL of distilled water was added to the tubes and the absorbances of the assay solutions were measured at 530 nm using a Biochrom 4060 UV-visible spectrophotometer (Pharmacia Biotech, Sweden). The ratio of the absorbance values for the samples and that for the test reaction was then used to calculate the volumetric β-glucanase activity in the sample by a method based on the enzyme kinetics for a batch reaction system (Beshay et al., 2007a). This method also allowed the estimation of the concentration of the target enzyme in the sample according to the kinetics of the reaction which further led to the parameter called selectivity for the target product ($S_{P/X}$) which was defined as the ratio of the concentration of recombinant enzyme (g

L^{-1}) to the dry biomass concentration (g L^{-1}). This term signified the ratio of the desired recombinant product to the biomass produced.

4.7.4 Total excreted protein content

Protein content in the culture supernatant, referred to here as extracellular total protein concentration was measured by the modified Bradford assay using the Roti®-Nanoquant kit (Carl Roth GmbH + Co. KG, Germany) by following the manufacturer's instructions. An acidic solution of the dye coomassie brilliant blue G-250 was incubated with an appropriate dilution of the supernatant. Bovine serum albumin (Albumin Fraction V; Carl Roth GmbH, Germany) was used to create a standard curve (Appendix, Table 8.2). The binding of the protein causes the transition of the dye from its reddish brown to blue form and a shift of its absorption maximum from 465 nm to 590 nm. The range of the assay was improved by taking the ratio of the absorbances at 590 nm and 450 nm, measured by a Biochrom 4060 UV-visible spectrophotometer (Pharmacia Biotech, Sweden).

4.7.5 Extraction of periplasmic proteins

In order to isolate proteins from the periplasmic space, 1 mL of culture sample was centrifuged at 9000g for 10 min in a Sigma 1-15 microfuge and the pellet suspended in 500 μL of a solution containing 200 g L^{-1} sucrose, 0.2 M Tris and 0.1 M EDTA (pH 8). After incubating on ice for 20 min, the samples were centrifuged again at the same conditions as earlier. After removing the supernatants, the cell pellets were resuspended in 500 μL of a solution containing 5 mM $MgSO_4$ and 10 mM Tris (pH 8) and incubated on ice for 10 min. The spheroplasts were centrifuged down and the supernatant was recovered as the periplasmic fraction.

4.7.6 Extraction of cytoplasmic proteins

A simplified extraction procedure was performed for rapid testing of the presence of the target protein in soluble and insoluble cytoplasmic fractions by SDS-PAGE. Defined volumes of samples from cultivations were taken corresponding to 0.1 g mass of cells and centrifuged to obtain cell pellets. The supernatant was taken as the extracellular fraction and the cell pellets were resuspended in 0.9% saline solution. The cells were lysed in a sonicator (Branson Ultrasonics, USA) with output control set at 2 and duty cycle set at constant. A pulse for 30 s and pause for 30 s were given in a tandem fashion. This cycle was repeated 3 times and the tubes were maintained on ice for the complete period in order to avoid excess heating. The lysates were centrifuged at 13000g for 10 min to separate the inclusion bodies along with the cell debris. The supernatant was taken as the soluble fraction. To dissolve the IBs, 0.5 mL of 8 M Urea was added to the pellet and mixed by vortexing. The centrifugation was repeated, and the supernatant fraction contained the previously insoluble protein fraction.

4.7.7 Immobilized metal ion affinity chromatography

The hexahistidine tag at the carboxy terminal of the recombinant β-glucanase enabled the use of immobilized metal ion affinity chromatography (IMAC) for the purification from the extracellular medium. An ÄKTAprime plus chromatography system (GE Healthcare Europe GmbH, Germany) equipped with a XK26 column including a borosilicate glass chromatographic tube and adapter AK26 (Pharmacia Biotech, Sweden) was used for the purification. Sepharose, a tradename of GE Life Sciences that stands for Separation Pharmacia Agarose, is a separation medium based on beads of highly cross-linked agarose with a mean diameter of approximately 90 μm functionalized with nitrilotriacetic acid (NTA) as a chelating ligand. Sepharose beads with immobilized NTA were charged with 100 mL of a $NiSO_4$ solution that had a concentration of 14 g L^{-1}. The column (50 mL bead volume) was washed with binding buffer (Table 4.11) over line A of the chromatography system at a flow rate of 3 mL min^{-1}. When the conductivity at the outlet reached a constant level, the column was assumed to be equilibrated and the UV_{280} detector was set to zero. Fermentation broth was centrifuged at 4000g for 10 min in a Sigma 6K15 centrifuge and the recovered supernatant was pumped through the chromatography system. The binding buffer was fed again into the system over line A until the UV_{280} reading in the outlet declined and remained at a steady level. The elution buffer (Table 4.12) was delivered through line B and a gradient elution was performed. The optimized procedure was to shift from 100% binding buffer (line A) to 100% elution buffer (line B) within a total volume of 50 mL. After that, elution buffer was continued to be pumped at 100% for another 20 mL. This corresponded to elution with a gradient of imidazole from 30 mM to 500 mM. The progress of the chromatography was viewed and the process data automatically saved by the associated software PrimeView 5.0 (Amersham Biosciences, Sweden). The eluted fractions were collected manually in volumes of 10 mL and analyzed by SDS-PAGE. After the chromatography, the nickel ions were removed from the column using stripping buffer (Table 4.13).

4.7.8 Protein precipitation by trichloroacetic acid

Proteins expressed at a low level in the extracellular medium were concentrated by trichloroacetic acid (TCA) precipitation for better visualization in SDS-PAGE. Culture samples (2 mL) from fermentations were centrifuged at 12000g for 10 min and the supernatants were recovered. To 1 mL of culture supernatant containing the dilute protein, 250 μL of a TCA stock solution with a concentration of 1000 g L^{-1} was added. The tubes were incubated on ice for 10 min and centrifuged at 12000g for 5 min at 4 °C in a microcentrifuge (Eppendorf AG, Germany). The supernatants were removed and the protein pellets were washed with 200 μL of ice-cold ethanol. The tubes were centrifuged again at 12000g for 5 min at 4 °C and the ethanol wash step repeated. After centrifugation and removal of the supernatant, the pellets were dried to remove the ethanol. The dried protein pellets were dissolved in 2x Laemmli buffer (Table 4.17) and incubated at 95 °C in a boiling water bath for 10 min in order to prepare the samples for loading onto a SDS-polyacrylamide gel.

4.7.9 Sodium dodecyl sulphate-polyacrylamide gel electrophoresis

Separation of denatured proteins on sodium dodecyl sulphate-polyacrylamide gel electrophoresis (SDS-PAGE) was performed based on the discontinuous pH principle by Laemmli (1970). The glass plates for preparing the cast were set together with a rubber seal and metal clamps. The separating and stacking gels were prepared according to the composition given in Table 4.27.

Table 4.27: Compositions of separating and stacking gel solutions for SDS-PAGE. Percentages refer to g of substance in 100 mL of solution.

Separating gel (12% acrylamide)		Stacking gel (5% acrylamide)	
Component	Volume (mL)	Component	Volume (mL)
Distilled water	1.5	Distilled water	0.775
1 M Tris pH 8.8	2.8	0.25 M Tris pH 6.8	1.25
Bis/Acrylamide (0.8%, 30%)	3.0	Bis/Acrylamide (0.8%, 30%)	0.425
5% SDS	0.15	5% SDS	0.05
10% Ammonium persulfate	0.0375	10% Ammonium persulfate	0.025
TEMED	0.0025	TEMED	0.0025
Total	**7.49**	**Total**	**2.527**

The separating gel was poured into the mould, and covered with a layer of 150 µL of isopropanol and allowed to polymerize. The isopropanol was removed and the gel surface washed with water. The stacking gel was poured onto the separating gel layer. The comb for forming the cavities was smeared with 5 µL of TEMED and set into the cast. After the polymerization was complete, the comb was removed and the cavities were washed with Tris-glycine buffer (Table 4.16). After removing the clamps and the rubber seal, any air bubbles present underneath the gel were removed by spraying a jet of water. The gel chamber was placed in an electrophoresis buffer tank which was subsequently filled up with Tris-glycine buffer. The electrodes of the chamber were connected to a current source Power Pack P 20 (Biometra GmbH, Germany).

Protein samples of 15 µL were diluted 1:1 with 2x Laemmli sample loading buffer (Table 4.17), incubated for 5 min at 96 °C and spun down at $12000g$ for 1 min before loading onto the gel. A pre-stained standard protein ladder SM0671 (Thermo Scientific, Germany) was loaded directly into the gel for comparison of molecular masses. The electrophoresis was started at a current of 9 mA and later increased to 14 mA when the sample front reached the separating gel. After the protein front reached the bottom of the gel, the electrophoresis was stopped and the gel was stained overnight in a coomassie brilliant blue G-250 solution (Table 4.18) and subsequently treated with the destaining solution (Table 4.19) in order to clearly visualize the bands. The gel was scanned and converted to a digital image using a HP Scanjet 5500c scanner (Hewlett-Packard GmbH, Germany).

4.7.10 Analysis of plasmid stability

To test plasmid segregational stability a sample of the culture was serially diluted up to 10^5 or 10^6 times depending on the OD_{600} such that about 100 CFUs were obtained upon plating a volume of 100 µL. The dilutions were first plated onto agar plates without antibiotics. After overnight growth, 70 colonies were picked and streaked onto plates containing 100 µg mL⁻¹ ampicillin. The ratio of the number of colonies that grew on selective medium to the number originally picked up from non-selective medium gave an estimate of the plasmid-carrying fraction in the population and thus the segregational stability.

For measuring the change in the plasmid content in the cells over time or over different growth rates, equal volumes (2 mL) of the cultures from fermentation were taken for the isolation of plasmid DNA using the kit Wizard® *Plus* SV Minipreps DNA Purification System (Promega GmbH, Germany). In the final step, the plasmids were eluted from the column into the same volume (50 µL) of elution buffer. The DNA concentrations in the preparations were measured using the NanoDrop® spectrophotometer. The final values were given as the isolated plasmid concentration normalized to the optical density of the original culture sample.

4.8 RNA Analysis

4.8.1 Precautions against RNase contamination

Since contamination by RNases could seriously endanger the integrity of RNA-based experiments and due to the ubiquitous nature of RNases in the environment, special precautions and extreme care had to be taken to ensure an RNase-free work space. Instruments like pipettes, electrophoresis equipment and working areas were cleaned with the RNaseZap solution (Life Technologies GmbH, Germany). Large surfaces were cleaned with 0.1 N NaOH. Plasticware for electrophoresis like gel tray, comb, *etc.* were cleaned with 0.1 N NaOH and 1 mM EDTA followed by rinsing with water that had been treated with diethylpyrocarbonate (DEPC). This compound inactivates RNases by covalent modification of histidine residues. Solutions were made RNase-free by treating with 0.1% DEPC, incubating overnight at 37 °C and autoclaving to remove DEPC. Exceptions to this procedure were Tris-based solutions since the amine group would interfere with DEPC. Such solutions were prepared with water previously treated with DEPC. Glassware was baked at 180 °C for 3 h to render it enzyme-free.

4.8.2 RNA protection and isolation

The RNeasy Mini Kit (QIAGEN GmbH, Germany) - a silica membrane spin column-based method - was used for isolation of total RNA longer than 200 bases from bacterial cells. The kit consisted of buffer RLT for cell lysis and RNase inactivation, buffers RW1 and RPE for column washing, and RNase-free water for elution. According to the manufacturer's instructions, about 100 µg RNA was the upper limit that could be isolated using 1 RNeasy mini spin column. For *E. coli* grown in minimal medium, this set a maximum of 7.5×10^8 cells that could be added to 1 mini spin column. Addition of more cells could decrease the efficiency of the isolation procedure as well as the purity of the isolated RNA. Based on plating experiments, a sample with an optical density OD_{600} of 1.0 contained a cell number concentration of 10^9 mL⁻¹, and thus, sample sizes were adjusted according to their optical densities. This was essential to ensure an equal RNA isolation efficiency for all samples. The RNeasy Mini Kit was reported to be efficient for RNA isolation for this scale of concentration of bacterial cells (Werbrouck *et al.*, 2007).

The RNAprotect Bacteria Reagent (QIAGEN GmbH, Germany) contained tetradecyltrimethylammonium bromide which ensured the rapid stabilization of RNA molecules within bacterial cells and helped in safely storing samples for later analysis. By mixing the cells with the reagent directly during sampling, changes in the RNA levels of samples during the experimental procedure could be avoided. To 2 volumes of RNAprotect Bacteria Reagent, 1 volume of bacterial cell culture needed to be added and mixed. If the sample would be drawn and stored on ice and the necessary volume of culture taken and added to the reagent (according to optical density) the sample would no longer be fresh and the transcriptome may have undergone changes during the incubation on ice. Therefore, a fixed volume (1 mL) of reagent was aliquoted to 2 mL RNase-free tubes and a fixed volume of bacterial cell culture (0.5 mL) was drawn into the reagent directly from the reactor. During the incubation of this mixture at room temperature for 5 min (according to manufacturer's instructions), a parallel sample was taken and the optical density was measured. Thus, the amount of cells present in the cell-reagent mixture was known and the volume of this mixture corresponding to 7×10^8 cells was taken for centrifugation. The tubes were centrifuged for 10 min at $5000g$. The supernatant was decanted and the cell pellet stored at -20 °C until later for RNA isolation. Thus, sampling was carried out in a manner that the cells were not subjected to any changes before mixing with the reagent while still being normalized according to their optical density.

Cells were lysed according to the manufacturer's guidelines under Protocol 1 (Enzymatic lysis of bacteria) and total RNA was isolated according to Protocol 7 (Purification of total RNA from bacterial lysate using the RNeasy Mini Kit). Cell pellets were thawed and 200 µL of TE buffer-lysozyme mixture (Table 4.20) was added and the pellets resuspended by careful pipetting. The mixture was vortexed for 10 s and incubated at room temperature for 10 min. During this period intermittent vortexing for 10 s each was carried out every 2 min. 700 µL of buffer RLT was added to the tubes and vortexed vigorously. Buffer RLT, according to the Material Safety Data Sheet from Qiagen, contained a high concentration of the chaotropic agent guanidinium thiocyanate. Together with the β-mercaptoethanol added extra to this buffer, this step helped in irreversibly cleaving disulphide linkages by reduction and thus denaturing RNases released during cell lysis. Any particulate matter formed was centrifuged down. The supernatant was taken in a fresh tube and 500 µL of 100% ethanol was added to aid the binding of RNA to the columns in the subsequent steps. The tubes were mixed by shaking vigorously and up to 700 µL of the sample was transferred to an RNeasy mini spin column placed on a 2 mL collection tube. The columns were centrifuged for 15 s at a centrifugal acceleration of $8000g$.

At this point, steps for on-column DNase digestion using the RNase-free DNase set (QIAGEN GmbH, Germany) for eliminaton of genomic DNA contamination were integrated into the procedure. The flow-through was discarded and 350 µL of buffer RW1 (which contained the chaotropic agent guanidinium thiocyanate and ethanol) was added to each column and the centrifugation repeated under the same conditions as earlier. For one spin column, 70 µL of buffer RDD and 10 µL of DNase I stock solution were mixed together gently and the resulting mixture added directly onto the column. The columns were incubated at room temperature for 30 min before the DNase I mix was washed away with 350 µL of buffer RW1 and centrifugation under similar conditions.

500 µL of buffer RPE (containing added 100 % ethanol) was added to the columns, and the centrifugation was repeated. The flow-through was discarded and the wash repeated for a duration of 2 min. Finally, the columns were placed in RNase-free 1.5 mL collection tubes and the RNA was eluted with 50 µL of RNase-free water by centrifugation at 8000g for 1 min.

For the off-column DNase treatment, freshly isolated RNA preparations were mixed with 10 µL of buffer RDD, 2.5 µL of RNase-free DNase I (both from Qiagen GmbH, Germany) and the reactions were made up to 100 µL with RNase-free water and incubated at room temperature for 30 min. The RNA from the reactions was finally purified by a silica membrane-based RNA clean-up procedure using the GeneJET RNA Cleanup and Concentration Micro Kit (Fisher Scientific - Germany GmbH) according to the instructions from the manufacturer.

4.8.3 RNA Quality control

The total RNA isolated from each sample using the RNeasy Mini Kit were analysed for quality on the basis of 4 parameters: concentration, purity, integrity and absence of DNA contamination.

The concentration and purity (absorbance ratios) were measured using the NanoDrop® 1000 spectrophotometer by using the RNA measurement mode. Absorbance at 280 nm is caused by protein contamination and pure RNA showed a unique absorbance peak at 260 nm only. Thus the ratio A_{260}/A_{280} was helpful in following this contamination. For pure RNA preparations, this ratio is supposed to be within the range of 1.9 to 2.1. The second ratio A_{260}/A_{230}, characterized the presence of other contaminants like guanidinium thiocynate or phenol and should also ideally be close to 2.0 for pure RNA preparations, but this ratio could drop significantly in case of low-concentration RNA preparations.

The integrity of the isolated RNA was checked by electrophoresis in a 2 % non-denaturing agarose gel in 1x Tris-borate EDTA buffer (TBE). The buffer (Table 4.22) was prepared separately using DEPC-treated water to minimize the risk of RNases. The RNA preparations were mixed in the ratio 1:1 with the 2x RNA loading dye (R0641; Fisher Scientific - Germany GmbH) and heated at 70 °C for 10 min. The mixtures were cooled on ice and spun down briefly before being loaded on to the agarose gel. The presence of 95 % formamide in the RNA loading dye ensured that RNA molecules were resolved to a certain extent even under non-denaturing conditions. The dye Roti-Safe® GelStain (Carl Roth GmbH + Co. KG, Germany) was added to the molten agarose gel according to manufacturer's instructions. As standard marker, RiboRuler Low Range RNA Ladder (SM1831; Fisher Scientic - Germany GmbH, Germany) was prepared in the same way as the RNA samples and loaded onto the gel.

Contaminating DNA molecules offered extra copies of the target which could seriously affect the reliability of the quantification of mRNA transcripts. In order to test for the absence of DNA contamination in the isolated RNA, the primers for amplifying the target region in the *rpoD* coding sequence (Table 4.3) were used in a qPCR experiment using the Rotor Gene SYBR Green PCR kit (QIAGEN GmbH, Germany) in the absence of reverse transcriptase (no-RT control). Pure RNA preparations should therefore ideally have a C_T matching that of a no-template control (NTC) the fluorescence of which was purely due to interaction and amplification of primer dimers. As positive controls, genomic DNA preparations were used with concentrations of 549 ng μL^{-1}, 54.9 ng μL^{-1} and 5.49 ng μL^{-1} named Std1, Std2, and Std3 respectively.

4.8.4 Reverse transcription quantitative real-time PCR (RT-qPCR)

Primers were designed to amplify a 100-150 bp long region in the different genes under study. To prepare uniform plasmid-based standards for quantification of each target amplicon, the primers were also used in a conventional PCR using Phusion polymerase to create blunt ended amplification products. These products were cloned into the pJET blunt vector backbone using the pJET cloning kit (Fisher Scientific - Germany GmbH). The clones were used to isolate fresh pJET-amplicon plasmids which were then quantified using the NanoDrop® spectrophotometer. Since 1 mol of the plasmid contained 1 mol of the cloned target amplicon, the following formula could be applied to determine the number of copies of any target amplicon per μL volume.

$$\text{amplicon number density (copies } \mu L^{-1}) = \frac{\text{Concentration of DNA (ng } \mu L^{-1}) \times 10^{-9}}{650 \text{ Da/bp} \times \text{Plasmid size (bp)}} \times 6.023 \times 10^{23}$$

A serial dilution of these purified plasmids and subsequent qPCR with known copies of target added to each reaction, allowed the construction of standard curves for each target amplicon. This gave a relation between the concentration of copies per reaction (C_r) and the threshold fluorescence cycle (C_T). The standard curve could then be used for the calculation of the amount of transcripts present in the samples using absolute quantification.

One-step reverse transcription and quantitative real-time PCR was performed using the Rotor Gene SYBR Green RT-PCR kit (QIAGEN GmbH, Germany). The kit included a master mix that contained a Hot-start Taq DNA polymerase, a buffer optimized for both reverse transcription and DNA polymerization, SYBR Green I dye for intercalating double stranded DNA, and a deoxynucleoside triphosphate (dNTP) mixture. The kit also included the Rotor-Gene RT Mix which was an optimized combination of the proprietary reverse transcriptases Omniscript® and Sensiscript®. A constant amount of starting RNA template from each sample was added to each reaction for a normalized analysis.

Since reverse transcription and subsequent PCR would be carried out in a single tube, only specific primers could be used for the reverse transcription. These were the reverse primers used for the PCR step (right primers in Table 4.3 with the exception of bla). The following reaction mixture (Table 4.28), excluding the template, was set up on ice as a cocktail and distributed to the corresponding number of 0.2 mL PCR reaction tubes. The corresponding template RNA was finally added to each tube.

Table 4.28: Reaction set-up for one-step RT-qPCR.

Component	Stock concentration	Volume taken (µL)	Final concentration
2x Rotor-Gene SYBR Green RT-PCR Master Mix	2x	12.5	1x
Left primer	50 µM	0.5	1 µM
Right primer	50 µM	0.5	1 µM
Rotor-Gene RT Mix		0.25	
Template RNA	Variable	Variable	80 ng per reaction
RNase-free water		Variable	
Total reaction volume		**25**	

The tubes were arranged in the 36-well rotor of a Rotor Gene RG-3000 real-time PCR instrument (Corbett Research, Australia) and the following 2-step cycling protocol (Table 4.29) was setup to run the experiment using the Rotor Gene software Version 6.1 Build 71.

Table 4.29: 2-step cycling protocol for one-step RT-qPCR

Step		Time	Temperature (°C)
Reverse transcription		10 min	55
Polymerase activation		5 min	95
Cycling (40x)	Denaturation	5 s	95
	Combined annealing and extension	10 s	60*

*Fluorescence data collection

The fluorescence data collection was performed at the end of each cycle after the extension step through the FAM/Sybr channel preset in the instrument by the manufacturer with the following settings: excitation at 470 nm, detection at 510 nm, gain factor 5. At the end of the real-time PCR cycling, a melting curve analysis was performed to verify the specificity of the amplicons produced and to check for any amplified side-products which could cause serious errors in the calculation of transcript amounts. In this procedure, the temperature was sequentially increased from 65 °C to 95 °C in steps of 0.5 °C per step while the fluorescence data were collected. The negative derivative of fluorescence with respect to temperature (-dF/dT) was plotted against the temperature in °C to determine the exact temperature at which the degradation of the amplicons in a reaction tube occured marked by a sudden peak due to the sharp decline in fluorescence. Multiple peaks from a melting curve analysis strongly suggested non-specific amplification. In consequence, the reaction mixture was electrophoresed on an agarose gel to further verify the amplification products.

4.9 Relative plasmid abundance measurement

The relative plasmid abundance in cells was calculated using quantitative real-time PCR (qPCR) based on the method by Lee *et al.*, 2006. The *rpoS* gene was chosen as the reference gene from the genome and the *bla* gene coding for β-lactamase was chosen as the target gene since it occurs as a single copy on the plasmid p582. Culture samples from continuous cultures at different space velocities were diluted to an optical density of 0.2 to normalize the amount of total DNA used in a reaction for each sample. The diluted cell samples were then lysed in a water bath at 98 °C for 5 min. The lysates were cooled gradually and taken as template total DNA (genomic and plasmid). Forward and reverse primers denoted as left and right for the respective targets *bla* and *rpoS* are given under Table 4.3. qPCR was carried out using the Rotor-Gene SYBR Green PCR Kit (QIAGEN GmbH, Germany) according to manufacturer's instructions for reaction setup and two-step cycling. The reaction set up was similar to the one in Table 4.28 with the exception that the master mix corresponding to qPCR was used and the reverse transcriptase was omitted. The cycling conditions were similar to the ones given in Table 4.29 with only the reverse transcription step being omitted. Standard curves for both *rpoS* and *bla* amplicons were generated using plasmids pJET-rpoS-frag. and p582 respectively and used for absolute quantification of the respective amplicons. The ratio of the calculated copies of *bla* transcript to the calculated copies of the *rpoS* transcript gave an estimate of the relative abundance of the plasmid in the cells with respect to the genome.

4.10 Fluorescence analysis

The activity of the stationary phase promoter P_{fic} was estimated by measuring the GFP fluorescence using the RF-5301PC spectrofluorophotometer (Shimadzu Corporation, Japan). Culture samples were diluted with water in the same manner as for the measurement of biomass by optical density. Therefore, a maximum OD_{600} of 1.0 was allowed and samples with a higher biomass concentration were diluted to this level. The software Panorama[2] Fluorescence (Labcognition Analytical Software GmbH & Co. KG, Germany) version 3.0.25.0 was used for analyzing the fluorescence. Under the menu command 'Photometric measurement', the following parameters were set: Sensitivity - low, Excitation slit width - 5 nm, Emission slit width - 5 nm. The fluorescence from GFP+-LVA was directly measured in the instrument with an excitation wavelength of 491 nm and an emission wavelength of 512 nm (*gfp+*; Scholz *et al.*, 2000). The instrument was blanked with water and for each sample, fluorescence for each sample measure in triplicate. Mean and standard deviation were calculated using Microsoft Excel 2010 (Microsoft Corporation, USA). Final fluorescence data were expressed as relative fluorescence per OD_{600} in order to normalize for the biomass concentration.

5 Results and Discussion

In the following sections, results from the various experiments would be described and further discussed in the context of information from the literature, in two major parts. The first part comprises of results from the establishment of a continuous cultivation process for extracellular recombinant protein expression which would be continued with the molecular genetic strategies pursued in order to develop an alternative antibiotic-free system. Following the characterization of these continuous cultivation systems and methods for their improvement, a closer look was taken on the genetic control over the extracellular expression, and the results from these experiments form the second part. The activity of the growth-phase regulated promoter which was responsible for the expression of BRP and hence excretion of the target protein into the medium during continuous cultivations in the first part was studied in greater detail using real-time PCR and transcriptional fusions. The aim was to deliver a closer picture of the control exerted by the cells at the molecular level during various specific growth rates in a chemostat.

5.1 Characterisation of growth and β-glucanase expression in *E. coli* JM109-p582

The starting point for the project was offered by the strain *E. coli* JM109 transformed with the plasmid p582 which was found to be the best performing strain in various media, both in terms of growth as well as recombinant protein expression (Spexard, 2007). For this reason, the combination of this strain and plasmid would henceforth be denoted as the reference system, in order to set a benchmark against which any future modifications would be measured.

5.1.1 Shake-flask cultivation

Initial experiments with the reference system *E. coli* JM109-p582 were carried out with the aim of characterization of the strain and to attempt reproducing the results reported by Spexard (2007). Firstly, at the shake-flask level, the strain showed to be capable of good growth and β-glucanase expression. The results shown in Fig. 5.1 point to a maximum biomass concentration of about 10 g L^{-1} achieved by 22.5 h of cultivation time after which stationary phase started to set in. The unexpectedly high biomass concentration was found to have resulted due to an increased amount of carbon source added during medium preparation. The initial glycerol concentration was found to be 15.1 g L^{-1} which gave a biomass yield coefficient ($Y_{X/S}$) of 0.66. Starting from 10 h cultivation time, the exponential phase of growth is quite evident from the increase in biomass concentration and decline in residual glycerol concentration. The decline in the ammonium sulphate concentration stopped at 2 g L^{-1} showing that during the stationary phase, the culture was limited by glycerol. The extracellular recombinant β-glucanase activity showed an initial increase during the growth phase and then starting from 20 h, after the stop of increase in biomass concentration, showed a further increase up to a maximum of 753.4 U mL^{-1} at 32 h of cultivation.

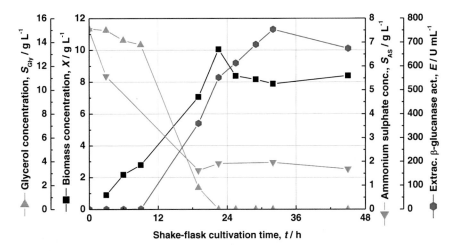

Fig. 5.1: Shake-flask cultivation of reference strain *E. coli* JM109-p582 in SGA medium at 37 °C. Culture conditions were as described in chapter 4.6.1.

5.1.2 Batch cultivation in a stirred fermenter

Following the shake-flask trial, the efficiency of the strain *E. coli* JM109-p582 in expressing the recombinant β-glucanase was tested in batch fermentation in the 7 L NLF 22 fermenter with 5 L working volume. As shown in Fig. 5.2, after 10 h of elapsed fermentation time the *DO* fell below the limit of 60 % and thus activated the impeller speed by means of which it was automatically maintained constant. The exhaust CO_2 analysed clearly mirrored the development of growth.

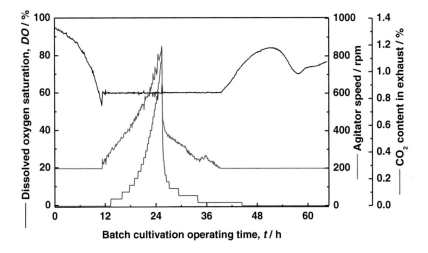

Fig. 5.2: Continuously monitored operating variables in batch process with reference strain *E. coli* JM109-p582. Culture conditions were according to those described for the NLF 22 fermenter in chapter 4.6.2.

The batch process gave a maximum biomass level of 7.2 g L^{-1} (Fig. 5.3). The slope of the curve for log of cell number plotted against time revealed a maximal specific growth rate of 0.24 h^{-1}. Further, a biomass yield coefficient ($Y_{X/S}$) of 0.72 with respect to glycerol was calculated. A maximum β-glucanase activity of 1.35 kU mL^{-1} during the stationary phase pointed to the activation of the P_{fic} promoter controlling the expression of the bacteriocin release protein that resulted in the spike in the extracellular enzyme activity during the onset of the stationary phase. A selectivity of 0.58 g g^{-1} was calculated for the production of the recombinant target enzyme (see chapter 4.7.3). A major fraction of the protein excreted by this strain is comprised of the target recombinant enzyme. Therefore, the extracellular total protein content measured by the modified Bradford method was found to be 1.33 g L^{-1} during the maximum production phase and it mirrored quite nearly, the profile of the recombinant enzyme activity.

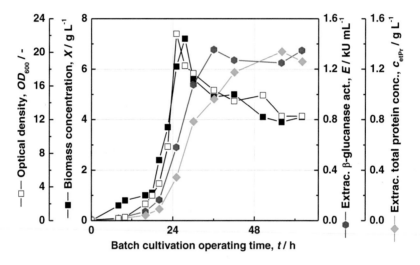

Fig. 5.3: Profiles of fermentation products during batch fermentation with the control strain *E. coli* JM109-p582.

The initial optimizations for the expression of this hybrid *bgl* gene under the control of its native *Bacillus* promoter had been carried out using the plasmid pLF3 in the strain *E. coli* JM109 (Miksch *et al.*, 1997b). The excretion of the product was by means of the BRP expressed by the *kil* gene under control of the stationary-phase promoter P_{fic}. Batch cultivations in TB medium with 5 L working volume yielded an extracellular enzyme activity of 150 U mL^{-1}. The next modification came in the form of placing the target gene under the control of the highly inducible T7 promoter along with zinc supplementation in a batch process whereby the activity of the product in the extracellular medium increased up to 2800 U mL^{-1}. Importantly, it was also shown that secretion to the periplasm was not a limiting factor and that cytoplasmic activities were low (Beshay *et al.*, 2007b). Optimizations involving differring strengths of promoter for the *kil* gene and the use of strong constitutive promoters for the target gene to avoid the use of expensive inducers, resulted in an extracellular β-glucanase activity of 91 U mL^{-1} (Beshay *et al.*, 2007c). The construction of the plasmid p582 followed by further optimizations and the use of additional devices containing Zn^{2+}-charged PDC metal-affinity resin for integrated production and purification resulted in an extracellular activity of about 500 U mL^{-1} for batch cultivation in TBG medium (Beshay *et al.*, 2009). This construct was further confirmed to be very efficient for β-glucanase production in batch and fed-batch processes yielding 703 U mL^{-1} and 4650 U mL^{-1} respectively (Spexard, 2007).

5.1.3 Fed-batch fermentation

Fed-batch cultivation, using a concentrated feed allows substrate limitation and control of the specific growth rate of the microorganism, and both of these parameters could be allowed to reach a quasi-steady state. The specific growth rate is sustained over an extended period resulting in a constant increase in cell concentration. In the strain *E. coli* JM109-p582, this was expected to be beneficial since an increased cell concentration would improve the volumetric enzyme productivity. More importantly, following an initial batch growth and accumulation of the recombinant enzyme, the low specific growth rate during the feeding phase could trigger a starvation response in the cells, which could activate the P$_{fic}$ promoter and promote excretion of the enzyme into the medium. Controlled administration of the substrate also aids in overcoming technical difficulties in oxygen transfer to the culture at high biomass concentrations (Enfors & Häggström, 2000).

The profiles of the important process parameters in the fed-batch fermentation of *E. coli* JM109-p582 are shown in Fig. 5.4. For feeding control, since the dissolved oxygen saturation in the medium is a function of the oxygen consumption rate of the culture which in turn could be controlled by the feeding rate, a coupling of the feed pump and the *DO* signal was possible. This offered the advantage that substrate would be fed only according to the rate at which it was actually consumed. A maximum limit of 60 % saturation was set on the *DO*, and therefore higher *DO* values activated the feed pump.

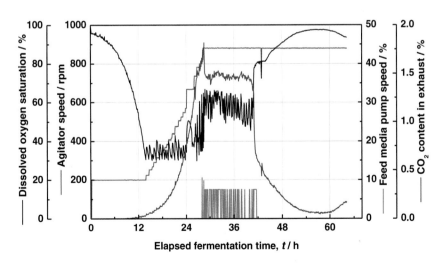

Fig. 5.4: Process data for fed-batch fermentation with reference strain *E. coli* JM109-p582. Culture and feeding conditions were as described in chapter 4.6.2.

The concentrated feed solution (see chapter 4.6.2) was fed at the rate of substrate consumption. This was seen from the observation that after decreasing for a while as a response to the substrate added, the DO increased again quickly. The CO_2 concentration in the exhaust air flow clearly contrasted with the batch process where the concentration fell sharply immediately after the growth phase whereas here, the CO_2 level was maintained below the maximum all through the feeding period pointing to biomass growth at low rates.

The feeding was successful in increasing the biomass further after the initial exponential growth phase, and a final dry biomass concentration of 14.6 g L^{-1} was reached (Fig. 5.5). The lowered specific growth rate during the feeding period (starting from 28 h) was deduced from the change in the slope of the plot of $\ln(OD_{600})$ versus time. From the glycerol consumption curve, it was seen that a quasi-steady state was reached during the feeding period (from 28 h) where there was no residual glycerol concentration and all the substrate fed was consumed instantaneously by the cells due to their high concentration. The profile for the residual ammonium sulphate concentration showed an upward trend after the onset of feeding while glycerol remained steadily below measureable levels pointing to a purely glycerol-limited condition.

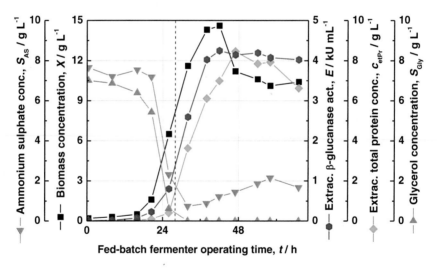

Fig. 5.5: Profiles of the concentrations of the main components during fed-batch fermentation of *E. coli* JM109-p582. The onset of feeding at 28 h is indicated by a dashed line. Culture conditions were according to the descriptions given in chapter 4.6.2.

The reference strain *E. coli* JM109-p582 yielded recombinant β-glucanase activity levels of up to 4 kU mL^{-1} from the fed-batch process which was thrice as high as the level achieved from the batch fermentation (Fig. 5.3). Taken together with the biomass concentration, this corresponded to a high selectivity of 0.9 g g^{-1} for the target protein. This could be explained by the fact that the cells were still growing during the feeding phase albeit at a lower rate, thus proving to be an optimal combination for the constitutive CP7 promoter controlling the *bgl* gene and the stationary phase P_{fic} promoter controlling the *kil* gene. This must have ensured that the periplasmic recombinant product concentration did not reach saturating levels and the product was always effectively transported to the medium. The extracellular recombinant enzyme profile was supported by the profile for total excreted protein which showed a very high concentration in the supernatant. Thus, it could be seen that the combination of constitutive promoter for the product gene expression until the end of the initial growth phase followed by the activation of the weak stationary phase promoter for the *kil* gene, was quite effective in ensuring high productivity and also transport to the extracellular space. Thus it was possible to reproduce the observations of Spexard (2007) and it could be stated that the reference strain *E. coli* JM109-p582 was working very well for batch and fed-batch modes of operation with chemically defined medium.

5.1.4 Purification by immobilized metal ion affinity chromatography

Following the successful expression of the recombinant β-glucanase in the extracellular medium, the His-tag present at the carboxy terminal of the protein could be exploited for a simple purification by immobilized metal ion affinity chromatography (IMAC). This would yield an enzyme preparation of high purity for future applications or further studies. To this effect, the final culture broth of the batch fermentation (see 5.1.2) was centrifuged to separate the cells. From the supernatant, 200 mL was loaded onto an IMAC column and chromatography was performed using the ÄKTAPrime system (described in chapter 4.7.7). After starting the elution with Buffer B (see Table 4.12), the fractions were collected in the following manner with respect to the elution volume: Fraction 1: 50 mL to 70 mL, Fraction 2: 70 mL to 100 mL, Fraction 3: 100 mL to 115 mL, Fraction 4: 115 mL to 140 mL, Fraction 5: 140 mL to 152 mL (Fig. 5.6). Immediately after the start of pumping Buffer B through the column, a peak in UV_{280} absorption could be seen which corresponded to the loosely bound host proteins that got eluted early. The fraction number 4 contained the target protein which was detected as a sharp peak in the UV_{280} absorption which occurred at 120.6 mL. This fraction was subjected to Bradford analysis which gave a protein concentration of 0.277 g L^{-1}.

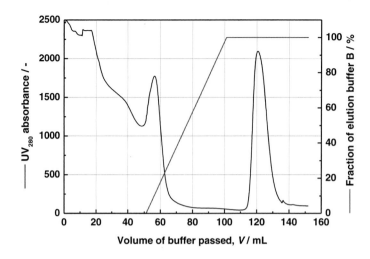

Fig. 5.6: IMAC chromatogram of supernatant of batch fermentation.

5.1.5 Electrophoresis of the separated protein aliquots

Culture supernatants from the stationary phase of the batch fermentation were analysed by SDS-PAGE and compared to the fractions purified by IMAC as shown in the gel photograph in Fig. 5.7. The lanes to the left of the standard protein marker (M) show unpurified supernatant samples from the batch fermentation where the target protein is seen along with the other host proteins although the target protein forms a major fraction of the total protein content excreted by the strain. The numbered lanes to the right of the marker show the different IMAC separated fractions. Most of the host proteins were removed in the first 2 aliquots along with a small fraction of the assumed target protein which constituted the loss during purification. The fraction number 4 which showed up as a clean peak in the chromatogram corresponded nicely on the SDS-PAGE gel as a dense band and this was the assumed pure β-glucanase protein. This overexpressed protein band occurred at a position corresponding to the 4th band below the orange-coloured 72 kDa standard. The corresponding molecular mass for the band correlated with the value of 26 kDa reported earlier (Beshay *et al.*, 2007a; Spexard, 2007). A small amount of loss was also seen in the fraction number 5 which was the tail of the pure protein peak in Fig. 5.6.

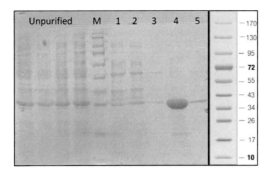

Fig. 5.7: SDS-PAGE analysis of IMAC aliquots. M refers to pre-stained protein ladder SM0671 which is also shown separately with corresponding molecular masses for comparison (Fermentas, Lithuania). The electrophoresis was carried out as described in Chapter 4.7.9.

For the expression of three different *Bacillus* enzymes in *E. coli*, Yamabhai *et al.* (2008) reported the best excretion into the culture medium, for the smallest of the enzymes mannanase with 41 kDa. This could help to explain the ease of excretion of the assumed β-glucanase of 26 kDa in this work.

5.1.6 Continuous cultivation in the absence of selection pressure

Having tested the reference strain under batch and fed-batch cultivation modes in defined medium, the next step was to establish a process in the continuous mode which would, when stable, enable the realization of higher enzyme productivity. The overall productivity would be affected by a collection of process as well as genetic parameters (Park *et al.*, 1990). The challenge here was to attain a steady state between the continuously dividing cells, *bgl* expression and maintenance of optimal growth rate to activate the *kil* gene at a steady but low level, so as not to affect the cell viability. Therefore, the most critical variable to optimize was the space velocity of the medium through the reactor (dilution rate) which had to ensure activation of the product expression as well as excretion.

A continuous culture was established in simple chemostat mode of operation with the control strain (*E. coli* JM109-p582). The working volume for the chemostat was 1 L and initially a space velocity of 0.1 h^{-1} was chosen, so as to be a safe measure away from the maximal specific growth rate of 0.24 h^{-1} (from the batch fermentation). Therefore, the feed inlet rate was 100 mL h^{-1} and the limiting substrate concentration (glycerol) of the inlet medium was kept at twice the batch process concentration, i.e. 20 g L^{-1} while all the other media components were maintained at the same concentration as in the batch process. At a feeding rate of 100 mL h^{-1} and culture volume of 1 L, the residence time was set to τ = $1/D = 10$ h. No kanamycin was added to the feed medium, hence there was no selection pressure. The aim was to characterize the plasmid segregational stability under these conditions. Only the initial batch medium contained kanamycin at the typical concentration (see chapter 4.3). When feeding started in the chemostat phase and culture broth was pumped out of the reactor vessel, the antibiotic concentration gradually decreased due to the washout effect. The isolated plasmid concentrations from culture samples normalized to their optical densities (at steady state) was followed through the experiment to determine the plasmid stability behavior (see chapter 4.7.10).

The batch process was allowed to run for 19 h and an OD_{600} of 16.7 was reached. The inlet and outlet pumps were then switched on. The feeding was initially set at a rate of 75 mL h^{-1} (D = 0.075 h^{-1}). After approximately 2 residence times (2×13 = 26 h) a steady state was reached with an average OD_{600} of 34 and an average steady state dry biomass concentration of 8 g L^{-1} (Fig. 5.8). After 117 h (9 residence times) in this space velocity, the medium flow rate was increased to 100 mL h^{-1} and the system was allowed to stabilize for 25 h before the next sampling. At this new steady state with D = 0.1 h^{-1}, the OD_{600} stabilized at an average value of 34.3 and average steady state dry weight concentration of 10.6 g L^{-1}. Since the carbon source concentration in the feed was 20 g L^{-1} an approximate steady state yield coefficient of 0.5 could be calculated for the biomass over the substrate. The substrate concentration curve showed that the residual glycerol in the medium (S_{Gly}) at steady state was always at near zero levels, revealing that the system was limited by glycerol feeding which was completely consumed at these flow rates. The permeabilized outer membranes of the cells due to the BRP expression may have caused a reduction in their ability to block or scatter light which manifested itself as a distinct reduction in the ratio of the OD_{600} to the dry biomass concentration.

The curve for the isolated plasmid concentration normalized to OD_{600} showed a clear effect of the absence of an active selection principle (Fig. 5.8). By the end of the process, the plasmid concentration isolated from the cells had reduced to about 10 % of the original value (15.06 ng μL^{-1} to 1.39 ng μL^{-1}). This pointed to the possibility that the culture had been gradually taken over by plasmid-free cells which had a growth advantage in the absence of intensive energy requiring processes such as the recombinant gene expression and the plasmid maintenance. The loss of the plasmid-bearing cell fraction was clearly mirrored in the extracellular recombinant enzyme activity, the value of which was seen to decline after a time point, although the cell concentration was maintained at almost the same level. A peak activity of 1748 U mL^{-1} was achieved during the process before it gradually declined to final values of around 46 U mL^{-1} near the end of the experiment at 194 h. Microscopic examination of the culture samples (400x magnification) ruled out the possibility of a contamination as the cause for the decline in enzyme activity.

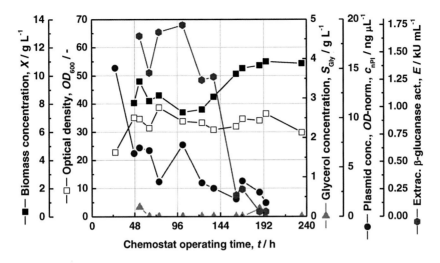

Fig. 5.8: Profiles of the main variables during a chemostat experiment with *E. coli* JM109-p582 in the absence of antibiotic selection pressure. The experiment was carried out in an in-house fermenter described in chapter 4.6.2 under culture conditions described in chapter 4.6.3.

5.1.7 Continuous cultivation in the presence of selection pressure

In this experiment, kanamycin was added to a final concentration of 50 µg mL^{-1} at one point during the process in the form of a sterile pulse injection, and the effects on the fraction of plasmid-bearing cells was followed by the isolated plasmid content and also substantiated by plating test. A space velocity of 0.1 h^{-1} was maintained throughout. The objective here was to find out if an intermediate addition of the antibiotic had a positive effect on the system in selecting for plasmid-bearing cells and improving the productivity and the results from this experiment are shown in Fig. 5.9.

The feed inlet and culture outlet pumps were started after an OD_{600} of 13.3 was achieved in batch mode (21 h from batch start). An average steady state OD_{600} of 23.9 and an average dry biomass concentration of 8.9 g L^{-1} were maintained up to 118 h chemostat time (140 h total process time), which was the point of antibiotic addition. After this point, the optical density was perturbed to a great extent. The later samples showed slightly varying biomass concentrations and cell densities. Nevertheless, the system was observed to run fairly stable for a total of 196.5 h chemostat time. The isolated plasmid concentration normalized to OD_{600} was found to be maintained at an average level of 6.1 ng µL^{-1} up till the end of the experiment. This showed that the antibiotic addition during the process, had helped in actively selecting for plasmid-bearing cells. The final value of the plasmid DNA concentration normalized to OD_{600} of the cell culture was 7.29 ng µL^{-1}, whereas this value was as low as 1.39 ng µL^{-1} in the experiment without antibiotic addition. It must be added however, the assumption here is that the plasmid content is equally distributed among the cell population.

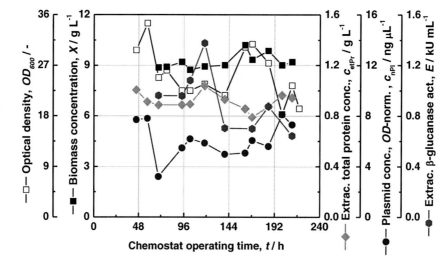

Fig. 5.9: Profiles of the main variables during chemostat operation with *E. coli* JM109-p582 in the presence of selection pressure due to kanamycin injected at 118 h. The experiment was carried out in an in-house fermenter described in chapter 4.6.2 under culture conditions described in chapter 4.6.3.

The extracellular β-glucanase activity was maintained at stable and high levels up till the end of the process. A comparison with the earlier trial without selection showed that the final specific enzyme activity was 69.9 U mg^{-1} cell dry mass against 4.18 U mg^{-1} cell dry mass in the former experiment. The extracellular total protein profile gave an average steady state protein concentration of 0.89 g L^{-1} which was maintained until the end of the process. Certain disturbances in the extracellular protein concentration towards the end of the experiment could have arisen as a result of lysis of non-growing cells due to the antibiotic addition. To sum up, it could be observed that the addition of antibiotic during the chemostat, helped in actively selecting for plasmid-bearing cells, resulting in a continued maintenance of productivity of the recombinant enzyme.

Figure 5.8 showed that when no antibiotic selection pressure was applied, a continuous decline in both plasmid content and extracellular enzyme activity occurred with respect to time. The volumetric enzyme productivity in that case declined from 131 kU L^{-1} h^{-1} at the point when dilution rate was set to 0.1 h^{-1} to 4.61 kU L^{-1} h^{-1} at the end of 195 h (96% decrease). The isolated plasmid concentration showed a clear effect of the absence of an active selection principle. However, Fig. 5.9 showed that it was possible to maintain both the plasmid concentration and the extracellular enzyme activity at relatively stable levels for long periods of continuous culture when external antibiotic selection pressure was applied. The maximum volumetric productivity in this case was 137.6 kU L^{-1} h^{-1} which dropped down to about 64.2 kU L^{-1} h^{-1} after 210 h process time (53% decrease).

Both quasi-lysis and shutdown of cellular protein synthesis are less severe upon expression of the ColE1 BRP in comparison to other BRPs like the colicin A lysis protein (Van der Wal *et al.*, 1995a; Cavard, 1991). This aspect could have helped in reducing lethality and for stably operating the chemostat with *kil* expression. Thus, apart from showing for the first time that continuous cultivation and extracellular product expression could be established in the reference strain *E. coli* JM109-p582, a dependence of plasmid stability on antibiotic selection pressure was also demonstrated as expected.

5.1.8 Continuous cultivation to find optimal operating condition

In order to study the growth and product expression kinetics with the reference strain *E. coli* JM109-p582 in greater detail, chemostat experiments were carried out at different space velocities under steady-state conditions at each space velocity. A higher space velocity would increase the specific growth rate ($\mu=D$), thereby increasing the output of biomass from the system ($D \cdot X$), decrease the steady state biomass concentration, and increase the residual substrate concentration. At a point near the critical dilution rate, when the maximum specific growth rate of the strain is exceeded, the biomass concentration should tend to zero and the residual substrate concentration gradually to the feeding concentration. At these substrate excess conditions, the stationary phase P_{fic} promoter is expected to be inactive. The effects of higher specific growth rate (μ) on the on the activity of the P_{fic} promoter and the ability of the strain to express the recombinant enzyme in the extracellular medium could be studied. If the BRP could still be expressed at a considerable level, it would help to increase the volumetric productivity of the recombinant enzyme from the system due to the high space velocity.

In this trial, the antibiotic was added already to the feed medium in order to ensure a steady supply (40 mL of kanamycin stock solution to 40 L feed) and prevent the disturbances in the steady state by intermediate antibiotic addition. The glycerol concentration in the feed was 20 g L^{-1}. One of the most significant observations from this experiment was that the biomass concentration could be stably maintained almost over the complete range of space velocities tested (Fig. 5.10 a). In fact, the profile for the biomass concentration could be seen to resemble the theoretical curves shown in Fig. 2.17 for a strain growing according to Monod kinetics.

The product formation kinetics were however quite different showing that expression and particularly excretion could not follow the profile of growth (Fig. 5.10 b). This led to steeply decreasing β-glucanase activities with increasing space velocity. An optimal condition may be found at space velocities around 0.2 h^{-1} with a maximal volumetric productivity of about 128 kU L^{-1} h^{-1}. The fall in the extracellular total protein concentrations was found to be slightly less drastic compared to the recombinant enzyme (also supported by the total protein productivity shown in Fig. 5.11). Interestingly, the concentration of plasmid isolated from the culture (normalized to optical density) showed an increasing trend with respect to the growth rate. However, it seems that this effect was not sufficient to compensate for a very slow rate of product release.

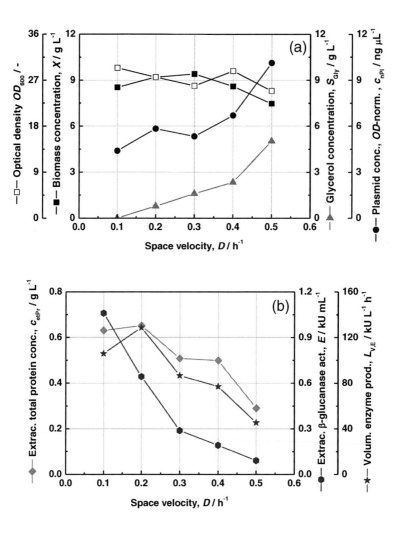

Fig. 5.10 Chemostat experiment with *E. coli* JM109-p582 at different space velocities. The experiment was carried out in an in-house fermenter described in chapter 4.6.2 under the general culture conditions described in chapter 4.6.3.

At the condition of lowest space velocity, the maximum selectivity ($S_{P/X}$) of 0.38 g g^{-1} was measured and it reduced to one-tenth of its value at the highest space velocity pointing to a clear shift in metabolism that reduced the level of target enzyme transported to the medium during high specific growth rates. The volumetric productivities of biomass and extracellular total protein were observed over the different flow rates and it was found that, the optimal value lay around 0.4 h^{-1} where the productivities peaked before falling off (Fig. 5.11). The biomass productivity showed a pattern which would have been expected for a strain growing with a kind of Monod kinetics. This is evident due to its similarity to the theoretical curve shown in Fig. 2.18 with the characteristic steady increase along with the space velocity resulting in a peak near the critical point and an eventual fall immediately thereafter.

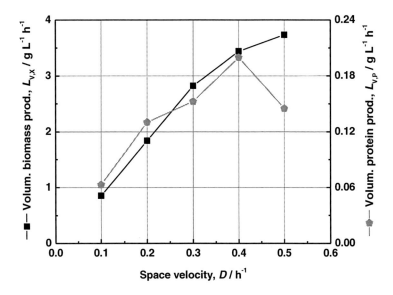

Fig. 5.11: Volumetric productivities in a chemostat with *E. coli* JM109-p582 at varied space velocities. The experiment was carried out in an in-house fermenter described in chapter 4.6.2 under the general culture conditions described in chapter 4.6.3.

Possible sources of experimental error that could skew steady state measurements were discussed by Herbert *et al.*, (1956). Firstly, the assumption of perfect mixing does not hold completely good in practice. Thus, there may have existed pockets in the culture where the dilution rates were less than the average and these pockets did not lose their biomass concentrations. The other important source of error, which was actually observable was the development of wall-growth. The adherence of layers of bacterial film to the reactor glass wall surfaces could not be completely prevented in this study and these layers may have acted as sources of continuous re-inoculation of the culture. The problem of wall-growth hindering accurate steady-state substrate measurements close to μ_{max} was also reported by Lendenmann *et al.* (2000).

In eqn. (5) the specific growth rate can be substituted with the space velocity for a process in steady state. This leads to the expression,

$$\frac{1}{D} = \frac{K_S}{\mu_{max}} \cdot \frac{1}{S} + \frac{1}{\mu_{max}} \tag{48}$$

By plotting $1/S$ versus $1/D$ for the data above, it was possible to calculate a K_S of 2.3 g L^{-1} for the reference system *E. coli* JM109-p582 in SGA for glycerol. Thus a saturation parameter s_{0S} of $20/2.3 = 8.69$ was used which correlated with the theoretical curve shown in Fig. 2.17. In retrospect, a higher feed glycerol concentration (S_0) could have been used which would have resulted in a favourably higher s_{0S} value and probably even better recombinant enzyme productivity.

The maximal specific growth rate of 0.24 h^{-1} observed in the batch fermentation acted as a guide towards the selection of the range of space velocities for this experiment, although from the observations in Fig. 5.10 it could be said that the strain growing at steady state in a chemostat displays significantly different growth characteristics as a batch culture and both the systems cannot be compared directly. In fact, the decrease in biomass concentration began at a space velocity of 0.5 h^{-1}. Similarly, in a study dealing with carbon-flux changes as a function of dilution rates in a glucose-limited chemostat culture of *Corynebacterium glutamicum* it was seen that the maximal specific growth rate observed at the batch scale could help in guiding the choice of dilution rates but need not correlate with the range of specific growth rates actually observed in the chemostat (Cocaign-Bousquet *et al.*, 1996). In fact, from the plot of $1/S$ versus $1/D$ described above based on equation (48), the y-intercept could be used to calculate the μ_{max} for the strain which in this case was 0.73 h^{-1}. This value correlates with the fall of biomass concentration beyond 0.5 h^{-1} as seen in Fig. 5.10a.

According to a study about the relation between membrane composition or fluidity and cell growth rate, an optimal space velocity of 0.3 h^{-1} that correlated with a significant increase in membrane phosphatidylglycerol was found to favour membrane fluidity and leakage of periplasmic proteins in *E. coli* (Shokri *et al.*, 2002). However, as shown in Fig. 5.10 in the present case, the activation of *kil* at low specific growth rates was very effective in bringing about membrane permeabilization and protein release. As part of the transition into stationary phase, overall rate of protein synthesis is reduced (Kolter *et al.*, 1993). This can also be expected to be the case during low growth rates in a chemostat.

The promoter used for the expression of the release protein has been shown to be crucial since an overexpression of *kil* could lead to cell lysis (Van der Wal *et al.*, 1995b). Previous studies using this system have shown the importance of modulating the promoter strength for the *kil* gene (Beshay *et al.*, 2007c). Though strategies using the type II secretion pathway and subsequent release of proteins accumulated in the periplasm either using the *kil* gene (Miksch *et al.*, 1997a; Robbens *et al.*, 1995; Aono, 1989) or using the Cloacin release protein (Lin *et al.*, 2001) have been known since many years, this study showed the possibility of taking the *kil* gene strategy forward to continuous and stable cultivation of *E. coli* with excretion of the recombinant product into the medium.

The use of a stable system for expression (constitutive expression of target gene plus moderately active *kil* gene) is vastly advantageous to other strategies (Hellmuth *et al.*, 1994) that use chemical or temperature-based induction mechanisms. The latter strategies make the measurements in terms of time post-induction and are inherently towards an unsteady state after induction. Adding to that there are problems of instability in both plasmid segregation as well as in expression.

The activity of promoter P_{fic} in continuous mode has been demonstrated and shows that growth phase regulated promoters could be used in continuous mode. This is because many functions identified to be invoked during stationary phase in batch growth may also be important while growing under nutrient-starved conditions. Therefore these functions can also be induced when cells are growing with a long doubling time, as for example in a chemostat at low space velocity (Kolter et al., 1993). The sigma factor gene rpoS for stationary phase is already activated at low growth rates near the stationary phase of a batch culture (Notley & Ferenci, 1996; Kolter et al., 1993).

The RpoS (σ^S) function does not require an absolute ceasing of growth and therefore carbon-limited chemostat cultures growing at low space velocities are capable of expressing increased RpoS activity. It was also suggested that this activity at low space velocities in a chemostat was even higher than the characteristic stationary phase activity in batch cultures (Notley & Ferenci, 1996). Several promoters induced during stationary phase show an inverse proportionality of expression with respect to growth rate (Kolter et al., 1993). It had been shown earlier that the level of the σ^S factor in the cells could be controlled by the dilution rate of the chemostat (Teich et al., 1999) and that decreasing dilution rates bring about increased RpoS levels (Zgurskaya et al., 1997). Thus, the optimal space velocity for the control culture E. coli JM109-p582 had to be a balance between the best growth rate, recombinant protein expression and the ideal level of substrate limitation that would activate the RpoS regulon. Since the P_{fic} promoter is known to be controlled by RpoS (Utsumi et al., 1993), the kil gene which is under the control of this promoter in p582 is expected to be activated by operating the chemostat at low space velocities.

5.2 Auxotrophic system based on leucine biosynthesis

The experiments described in Chapters 5.1.6 and 5.1.7 clearly showed that the strain E. coli JM109-p582 could be used for continuous cultivation and recombinant protein expression but the stability of the recombinant plasmid was dependent on the addition of antibiotic to the medium for an effective selection pressure. Though, this was able to demonstrate the plasmid stability, it is hardly a method of choice for an efficient long-term cultivation process. Therefore, the need arises to develop alternative antibiotic-free plasmid selection strategies. As mentioned in chapter 2.4, a variety of amino acid biosynthetic pathways have been targeted in the past in order to create an alternative selection pressure for the maintenance of plasmids in auxotrophic strains without the use of antibiotics. Although the leucine biosynthetic pathway has been known as a plasmid selection principle for experimentation with yeasts, its bacterial counterpart has so far only been studied with the leuD gene (Borsuk et al., 2007). The pathway however, holds promise for further investigation, in the form of other crucial genes that could bring about a selection pressure for a plasmid supplying the complementing copy of the gene. One such gene in the leucine biosynthetic operon is leuB coding for 3-isopropylmalate dehydrogenase and which forms the major focus of the experiments described below dealing with its adaptation into an alternative plasmid selection principle.

5.2.1 Cloning of *leuB* expression cassette onto plasmid p582

As shown in Fig. 2.7, for the cloning of *leuB*, it was not possible to use the natural promoter region without including the *leuA* gene along. Since there would always be a copy of *leuA* in the genome, having it on a plasmid, would increase the chance of recombination. So, the cloning strategy was forced to include an intermediate step during which the amplified *leuB* fragment was fused to another promoter before cloning into the target p582 plasmid. In fact, a similar strategy of sub-cloning to include a different promoter sequence was reported in the study involving the *leuD* gene (Borsuk *et al.*, 2007).

The complementation plasmid pFC2 was created by bringing the gene *leuB* under the control of a constitutive *Bacillus sp.* promoter P_{bgl} and cloning it at the unique EcoRI site on p582 between the *kil* and *bgl* genes. Firstly, the *leuB* coding sequence along with its ribosome binding site was PCR-amplified from genomic DNA of *E. coli* K12 MG1655 using primers 1 and 2 (Table 4.3). The PCR product was purified with Wizard SV Gel and PCR Clean-Up system (Promega GmbH, Germany) and restricted with XbaI (Fermentas, Lithuania) to generate an insert that contained a cohesive end on one side and a blunt end with an EcoRI site at the other end. Plasmid p55 which had been constructed by transferring the promoter region of the beta glucanase gene from *Bacillus amyloliquefaciens* on to pUC19 was suitable as an intermediate vector to borrow the P_{bgl} promoter. Plasmid p55 was cut with XbaI (cohesive) and HincII (blunt) before being subjected to 5'-phosphate dephosphorylation to generate pV55. The vector pV55 was ligated with the insert previously generated to result in p55-leuB. This plasmid now had the promoter P_{bgl} controlling the cloned *leuB* gene and the complete construct being flanked by two EcoRI sites. An EcoRI digestion and purification of insert resulted in the fragment 'Insert LeuB'. This fragment was ligated to p582 which had previously been cut at its unique EcoRI site and dephosphorylated. Colonies from Eckhardt gels that showed a CCC band that was larger in molecular mass compared to the corresponding band in the control were potential correct recombinants. The plasmids from these clones were isolated and restricted with EcoRI to check for the release of insert and the result of this screening is shown in Figure 5.12.

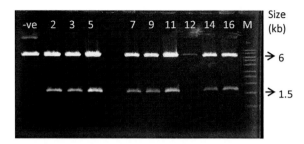

Fig. 5.12: EcoRI restriction screening for the presence of *leuB*. Plasmid from one of the colonies was used as a negative control (-ve) since the Eckhardt gel result had shown it to be a recircularized vector. The other preparations showed the release of the cloned insert with a size of around 1.6 kb. M refers to 1 kb molecular size marker (Plasmid Factory GmbH, Germany).

These positive clones were named as 'Final clones' series or FC. Colony 2 henceforth referred to as pFC2 was verified by sequencing and its plasmid map is shown in Fig. 5.13. Interestingly, sequencing of the remaining clones revealed that many of the other clones had the same orientation of the *leuB* gene as pFC2.

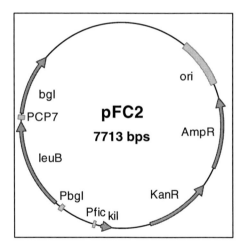

Fig. 5.13: Plasmid map of *leuB* complementation plasmid pFC2.

5.2.2 Complementation of *E. coli* JW5807-2 with construct pFC2

The Keio *ΔleuB* knockout strain *E. coli* JW5807-2 was made chemically competent and transformed with the construct pFC2. A shake flask experiment on SGA minimal medium with different conditions was carried out to test the auxotrophy of the strain and its complementation and the result is shown in Figure 5.14. The knockout strain was unable to grow on SGA medium while being able to grow normally when leucine was supplemented to the medium thus showing a marked dependence on added leucine ('Leu'). The knockout strain had been tested separately for growth on rich medium and had been found to be viable. Thus for the knockout strain in SGA, the presence of leucine was necessary for growth, shortened the lag phase, and higher concentrations of leucine increased the final biomass concentration as a result of leucine being used a carbon and nitrogen source. Leucine was tested at concentrations of 50 mg L^{-1} and 100 mg L^{-1}. Transformant colonies of *E. coli* JW5807-2-pFC2 however, grew on SGA minimal medium in the absence of leucine. The growth showed a longer lag phase but higher final biomass levels, ultimately proving that the plasmid was able to efficiently complement the leucine auxotrophy. Although the use of minimal medium was necessary for this auxotrophic system to be stable, there was freedom of experimenting with different carbon sources since this is an anabolism-based auxotrophy. The growth of the complemented strain in the presence of ampicillin ('Amp'; final concentration 100 mg L^{-1}) was an added evidence for the maintenance of the plasmid in the actively growing cells.

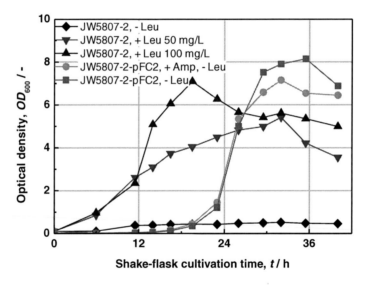

Fig. 5.14: Shake-flask cultivations of the knockout Keio strain *E. coli* JW5807-2 in the absence as well as in the presence of the complementing plasmid pFC2 in SGA minimal medium under different conditions. '+' and '-' refer to presence and absence of a compound respectively.

Separate shake-flask experiments carried out in order to calculate the maximal specific growth rate of the complemented auxotrophic strain *E. coli* JW5807-2-pFC2 in SGA medium, gave a value of 0.22 h^{-1}.

5.2.3 Inclusion bodies with the Keio strain

When observed under the microscope, samples from shake-flask cultivations at 37 °C of the auxotrophy-complemented strain *E. coli* JW5807-2-pFC2 showed unusually long undivided chains of cells towards the stationary phase of growth (see Fig. 5.16 A & B). Cells observed at 400x magnification showed a few long chains of undivided cells (A) whereas at 1000x magnification chains of cells with regularly spaced 'white spots' could be observed (B). This could have been due to extreme stress on the knockout cells that had to complement the leucine auxotrophy by maintaining the recombinant plasmid. The bright spots were suspected to be the overexpressed recombinant protein accumulated as inclusion bodies due to the prevailing stress conditions in the cells. Further support for this possibility was provided by the isolation of the insoluble protein fraction of cells from shake-flask cultures and their analysis by SDS-PAGE. The cells were lysed by ultrasonication and the cytoplasmic fraction was separated by centrifugation (see Chapter 4.7.6). The inclusion bodies were expected to settle down along with the cell debris and they were solubilized in 8 M urea, separated and analysed by SDS-PAGE. The result shown in Figure 5.15, suggested that the expressed recombinant protein could be locked up as inclusion bodies in the stressed cells since the urea-solubilized fraction gave a significantly strong band at the expected 26 kDa position which was absent in the extracts from the host strain lacking the plasmid.

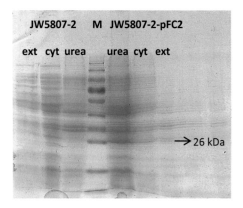

Fig. 5.15: SDS-PAGE analysis of cell-free supernatant (ext), cytoplasmic soluble fraction (cyt) and urea-solubilized fraction (urea) from shake-flask cultures of *E. coli* JW5807-2 with and without plasmid pFC2. The cultivations were carried out at 37 °C according to the descriptions given in Chapter 4.6.1.

The supposed band for the β-glucanase was clearly differentiated by its presence in the case of the strain with the plasmid pFC2 while being absent in the sample from the strain only. The relative band intensity was not very significant in the extracellular fraction whereas the urea solubilized fraction gave a strong band. If the expressed β-glucanase was in fact trapped in the form of inclusion bodies within the cell, it could explain the low extracellular enzyme activities that were observed with these strains, since the signal sequence was not available for targeting into the periplasm and hence no transport to the extracellular space could occur.

In order to minimize stress and inclusion body formation, the complemented strain *E. coli* JW5807-2-pFC2 was tested at lower cultivation temperatures of 25 °C or 30 °C which resulted in a slightly better recombinant β-glucanase activity in the case of 30 °C. Microscopic observations showed properly dividing cells and the absence of abnormal undivided chains of cells (Fig. 5.16 C).

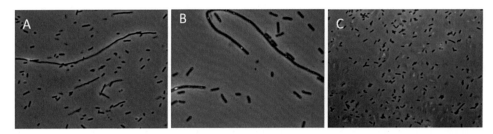

Fig. 5.16: Microscopic observation of the samples from the culture *E. coli* JW5807-2-pFC2 grown in shake-flasks in SGA minimal medium without leucine at an incubation temperature of 37 °C (A and B) or 25 °C (C). The cultivations were carried out according to the descriptions given in Chapter 4.6.1.

The effect of process temperature for cultivations with recombinant *E. coli* was recently studied and both recombinant antibody fragments (Fab) formation and release were reported to be inversely proportional to the growth temperature while having no effect on the biomass yield (Rodríguez-Carmona *et al.*, 2012). Due to these reasons, a process temperature of 30 °C was fixed in this work as standard for cultivation of auxotrophic strains in SGA medium.

5.2.4 Batch fermentation of *E. coli* JW5807-2-pFC2

A batch fermentation was carried out in the 2 L in-house fermenter with a working volume of 1 L in SGA minimal medium containing no leucine and no antibiotics. The inoculum was taken from a 24 h culture also grown in SGA medium and a reduced cultivation temperature of 30 °C was maintained (Fig. 5.17). A biomass concentration of 8.1 g L^{-1} was reached after 34.5 h process time. The maximum extracellular β-glucanase activity was measured to be 509.3 U mL^{-1} which was clearly not as high as that achieved with the control strain. This difference could probably be attributed to the different strain and plasmid used and the fact that the cells were also dependent on expression of the plasmid-based complementation gene for survival. As expected, the selectivity of the target protein with respect to biomass reduced to 0.19 g g^{-1} for this fermentation whereas, the selectivity was 0.58 g g^{-1} for the reference strain (see chapter 5.1.2).

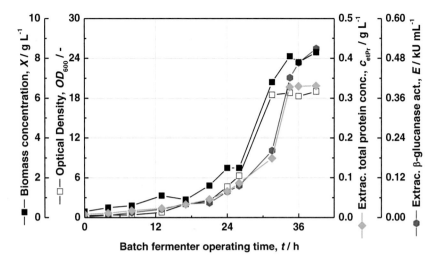

Fig. 5.17: Profiles of fermentation products during batch fermentation with *E. coli* JW5807-2-pFC2. The fermentation was carried out in the in-house fermenter according to the descriptions in Chapter 4.6.2.

A correlation between extracellular enzyme activity (kU mL^{-1}) and extracellular total protein concentration (g L^{-1}) could be deduced. In the reference strain *E. coli* JM109-p582, the ratio between these two parameters was found to be 1.07 kU mg^{-1}. Since the extracellular protein is chiefly dominated by the recombinant enzyme only, this ratio should not be deviating much from this level even when the enzyme activity was not very high. In the *E. coli* JW5807-2-pFC2 batch fermentation this ratio was 1.51 kU mg^{-1}.

The maximum cell concentration achieved with *E. coli* JW5807-2-pFC2 (OD_{600} 19.0) was comparable with the reference strain *E. coli* JM109-p582 (OD_{600} 22.2) but the maximal specific growth rate was about 6% lesser. This reduction in growth rate with complemented auxotrophic systems was also observed in a *Lactococcus lactis* auxotrophic complementation system based on the threonine biosynthetic pathway genes and is thought to be due to the necessity for endogenous amino acid synthesis under selective conditions (Glenting *et al.*, 2002). However this was very different from the *infA* complementation system for which no growth rate difference between control and plasmid-harbouring strains was to be found (Hägg *et al.*, 2004).

Supernatant samples from batch fermentation with the strain *E. coli* JW5807-2-pFC2 in a 5 L NLF fermenter were analysed on SDS-PAGE to check for the expression of the recombinant protein. For comparison, final supernatant sample from a shake-flask cultivation of the same strain is also shown in the Figure 5.18. The gradual increase in the band intensity of the recombinant protein in the extracellular medium could be clearly seen.

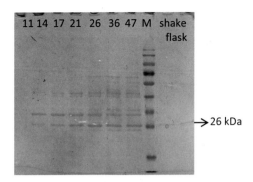

Fig. 5.18. : SDS-PAGE analysis of samples from 11h to 47 h from batch fermentation of strain *E. coli* JW5807-2-pFC2 in a culture volume of 5 L. The supposed β-glucanase band is seen at a position near 26 kDa. M refers to pre-stained protein ladder SM0671 (Fermentas, Lithuania).

5.2.5 Antibiotic-free continuous cultivation of *E. coli* JW5807-2-pFC2

Following the encouraging result with the batch mode of operation, a chemostat was set up in the in-house fermenter of 2 L total volume in which a working volume of 1 L was maintained. The feed medium contained 20 g L^{-1} glycerol as carbon source of which 40 L were prepared to last for a longer period of time. No antibiotics were used at any occasion during the process so that the efficiency of this complementation strategy as a means of plasmid selection could be verified. At the end of the batch phase, the chemostat was started by feeding at the rate of 0.1 L h^{-1}. This corresponded to a space velocity of 0.1 h^{-1} which was the same condition applied in the initial experiments using the reference strain *E. coli* JM109-p582 (see chapter 5.1.7). Therefore, this chemostat served to demonstrate the efficiency of the auxotrophic selection principle as a replacement to antibiotic based-selection on an extended time basis in terms of stability of the plasmid and the expression of recombinant enzyme (Fig. 5.19).

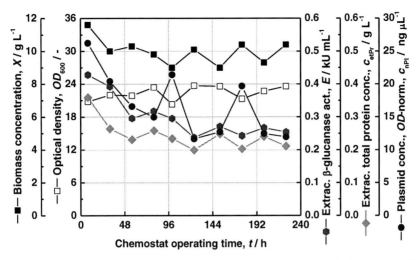

Fig. 5.19: Chemostat experiments with *E. coli* JW5807-2-pFC2 at a constant space velocity ($D = 0.1$ h^{-1}) and in the absence of antibiotics. The continuous cultivation was carried out at an incubation temperature of 30 °C in the in-house fermenter described in Chapter 4.6.2 with other experimental conditions as described under Chapter 4.6.3.

The graph showed that the auxotrophic system was very efficient in maintaining a stable production of the enzyme activity in the extracellular space without the aid of antibiotic selection pressure. The biomass concentration stabilized at an average of 10 g L^{-1} under these conditions. The plasmid concentration isolated from a fixed volume of the culture and normalized to the optical density (OD_{600}) showed a decline from 26.2 ng µL^{-1} to 11.9 ng µL^{-1} with a clear tendency to stabilize at the latter level (224 h). Although a fall of more than 50% was seen, the reduction had been as high as 90% in the reference strain over the same time period when no antibiotic selection pressure had been applied (Fig. 5.8). After the extracellular total protein concentration and the recombinant β-glucanase activity seemed to decline initially, they stabilized at average values of 0.245 g L^{-1} and 285.6 U mL^{-1} respectively.

During operating times of 96 h and 192 h, 9 and10 colonies respectively were chosen at random from the LB-ampicillin plate that was used for the replica plating for plasmid stability testing (see chapter 4.7.10). These colonies were grown overnight in culture tubes, their plasmids were isolated and analysed after EcoRI restriction. All of the clones contained the plasmid pFC2 in the right form and the released insert (above 1.5 kb) further confirmed the same (Figure 5.20). Since these plasmids were from single colonies, it was sure that the clones counted from the antibiotic plate were truly plasmid-carrying cells of the Keio strain and not any foreign contamination resistant to the antibiotic.

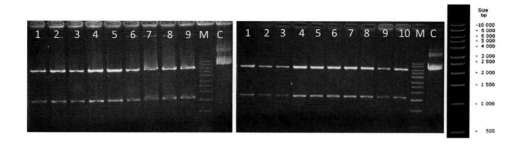

Fig. 5.20: Plasmids analysed by EcoRI restriction after being isolated from 9 and 10 colonies randomly chosen at 96 h (left) and 192 h (right) of chemostat operating time respectively. The last lane in each gel shows a control (C) where one of the plasmid preparations was loaded unrestricted. M refers to 1 kb molecular size ladder and is shown separately for comparison (Plasmid Factory GmbH).

In other trials it was found to be very difficult to create a non-selective environment for the *E. coli* JW5807-2-pFC2 culture by growing it in SGA medium containing 100 mg L^{-1} leucine to generate plasmid-free cells. This observation represents an advantage, since leucine can be added to the antibiotic-free medium, without the risk of a takeover by the plasmid-free cell population. The concentration of the leucine can be optimised for growth and recombinant protein production. The fact that this could still maintain the plasmid while removing some of the limitations of a completely minimal medium could possibly reduce stress and improve productivity. However, this has to be taken with caution, since in a chemostat, the presence of leucine may create a second source of limiting carbon substrate. This aspect of catabolism of the complementing amino acid as a preferred carbon source was also discussed in an auxotrophic system based on proline biosynthesis (Fiedler & Skerra, 2001). Also, the price of leucine is another important factor to be considered in this regard.

5.2.6 Testing different space velocities for continuous cultivation of *E. coli* JW5807-2-pFC2

To study the kinetics of growth and product release during continuous cultivation of *E. coli* JW5807-2-pFC2, different space velocities were explored and the resulting changes in cell concentration, extracellular β-glucanase activity, plasmid concentration and the segregational stability of the plasmids were tested and the results obtained are shown in Figure 5.21.

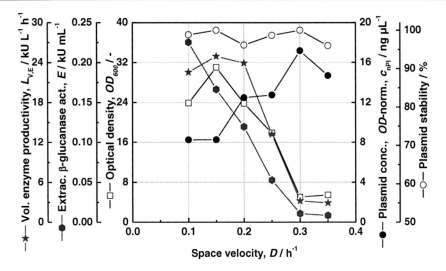

Fig. 5.21: Steady-state chemostat operation with *E. coli* JW5807-2-pFC2 at different space velocities and in the absence of antibiotics. The continuous cultivation was carried out at an incubation temperature of 30 °C in the in-house fermenter described in chapter 4.6.2 with other experimental conditions as described under chapter 4.6.3.

The optical density of the culture declined regularly at high flow rates although there were a few difficulties in maintaining the steady state at the highest space velocity which was close to the critical specific growth rate. The extracellular recombinant β-glucanase activity was also found to steadily decrease at higher space velocities mainly due to the associated decrease in biomass concentration. The decline seen in the optical density as well as the extracellular recombinant β-glucanase activity was almost linear as a function of the space velocity. This profile correlates with that of growth kinetics of first order with respect to a limiting substrate. In this case, the specific growth rate becomes a linear function of the limiting substrate concentration and represents the worst case of the simple Monod-type growth kinetics. In such a case for batch cultivation, the mass balance of biomass is represented by the so-called logistic function. Apart from a high K_S value, the use of the auxotrophic complementation system could have increased the maintenance costs in the strain which meant substrate was being consumed for cell maintenance rather than biomass. This could have resulted in the steep decline of steady state biomass concentrations when medium flow rate was increased. Thus it is evident that, the Monod kinetics that was seen for the reference strain in Figs. 5.10 and 5.11 was lost in the case of Fig. 5.21 mainly due to the different strain and plasmid used and perhaps also due to the effects on growth caused by the auxotrophy to be complemented.

The volumetric enzyme productivity showed a maximum of 25 kU L^{-1} h^{-1} at an optimal dilution rate of 0.15 h^{-1}. The plasmid concentration isolated from a uniform culture volume and normalized to cell density, showed a slight increase towards higher growth rates (Fig. 5.21), similar to the observation made with the reference strain (Fig. 5.10). The construct p582 and its derivatives were based on the pUC19 origin which maintained a higher copy number than pBR322 which is the frequent type of origin of replication reported in the literature (Lin-Chao & Bremer, 1986). Thus, the characteristics of plasmid copy number relation with the growth rate is expected to be different in this work, especially in the context of an auxotrophic selection mechanism that is itself coupled to the cell's metabolism. Recently, the plasmid content normalized to the biomass (mass of plasmid per biomass) has been compared over various growth rates and it shows a trend that is mixed to opposite to that observed in this work (Wunderlich *et al.*, 2014). However, the result of replica plating experiments showed a stable segregation of plasmids at near 100% (Fig. 5.21) for the entire period of the chemostat experiment (235 h).

The possibility of recombination to the genome has been postulated as a possible reason for loss of productivity in a process involving *Saccharomyces cerevisiae* (Meinander & Hahn-Hägerdal, 1997). However, in the construct pFC2, this risk is greatly minimized since the flanking regions to the *leuB* coding sequence on the plasmid are completely different to the natural sequences. The promoter used is a *Bacillus* β-glucanase promoter and the downstream sequences lead into the recombinant gene, whereas in the genome the operon continues into *leuC*. The authors also discuss the possibility of 'leakage' of leucine into the medium and uptake by other cells, causing a loss in selection pressure and even suggest that auxotrophic complementation of amino acids may not be a suitable strategy for long-term cultivations with *S. cerevisiae* (Meinander & Hahn-Hägerdal, 1997). However, in the experiments with continuous culture in this work using *leuB* as the selection principle for over 200 h of operation no loss in plasmid stability could be observed (Refer to Fig. 5.62).

The reason for lower enzyme productivity with the auxotrophic system in comparison with the reference system could be a result of either an increased burden due to a larger plasmid (p582-*leuB*) or more likely a result of the auxotrophic strain (Çakar *et al.*, 1999). On the other hand, the phenomenon of the complementing gene interfering with the productivity from the recombinant product is discussed in the study involving glycine as a selection principle (Vidal *et al.*, 2008), but could probably be ruled out in this work since the product gene (*bgl*) occurs downstream of the complementing gene (*leuB*) and is also driven by a relatively stronger constitutive promoter CP7 (Fig. 5.13).

5.2.7 Modification of plasmid pFC2 by removal of kanamycin resistance gene

Since the *leuB* auxotrophic strain *E. coli* JW5807-2 had a kanamycin resistance gene that had replaced the *leuB* gene, the kanamycin resistance gene on the complementation plasmid pFC2 was redundant and could be removed (Fig. 5.13). Even when antibiotic selection pressure would not be applied, it would be beneficial to remove the antibiotic resistance genes on a complementation plasmid (Peubez *et al.*, 2010). As a first step towards modification of plasmid pFC2, the kanamycin resistance gene was to be deleted (Knüttgen, 2013). This might bring additional benefits in the efficiency of gene expression due to reduced stress from the smaller sized plasmid apart from achieving one step further towards an antibiotic-independent system that is particulary attractive for the production of therapeutic and vaccine DNA due to regulatory requirements (Luke *et al.*, 2009).

Initially, the strategy was to use primers designed to amplify the complete pFC2 sequence other than the kanamycin resistance gene and circularize the product. After repeated failed attempts in generating the correct amplicon, it was thought that the extremely large size of the amplicon required in this case (6.9 kb) could be the limiting factor. Therefore, the strategy was changed to simply use restriction enzymes that cut near the ends of the kanamycin resistance gene portion, thus removing most of it (661 bases out of 815). The enzymes chosen for this purpose were Van91I and PspXI. These enzymes created cohesive ends and therefore the restricted DNA had to be treated with Klenow fragment for creating blunt ends. There was no necessity to phosphorylate the linear product since the ends had been generated from cuts within the DNA. Self-ligation was successful and transformation into *E. coli* Top 10 competent cells gave positive clones which were first analysed by restriction with EheI (also called SfoI) or HindIII (Fig. 5.22).

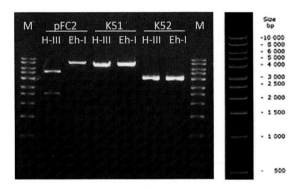

Fig. 5.22: Restriction analysis for screening of clones after kanamycin resistance gene deletion from pFC2. M refers to 1 kb molecular size marker which is also shown separately for comparison (Plasmid Factory GmbH, Germany). H-III and Eh-I refer to restriction enzymes HindIII and EheI respectively.

The original plasmid pFC2 had two sites for HindIII and one for EheI which is seen for the corresponding bands under pFC2 in Fig. 5.22. The next pair of lanes show the restricted bands from a clone K51. One of the HindIII sites which was present within the kanamycin resistance gene had been lost in the reduced plasmid resulting in a single band. The single restriction by EheI is shown not to be affected and the linear bands are seen to be of slightly lesser size compared to the single band in the lane pFC2 which could be due to the loss of 0.6 kb. Another clone K52 was probably the result of some error during the double digest step resulting in a smaller product. K51 was sequenced in the region of the kanamycin resistance gene and compared to a hypothetical sequence and verified to be correct. This clone was designated as pFKN (Fig. 5.23) and was stored as a glycerol stock in the strain collection of the Fermentation Engineering group.

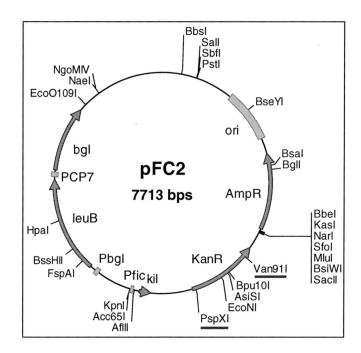

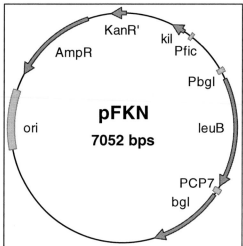

Fig. 5.23: (above) Construct pFC2 showing important restriction enzyme sites. The sites Van91I and PspXI have been highlighted. (below) The minimized construct pFKN showing the residual portion of the kanamycin resistance gene as KanR'.

5.2.8 Construction of a resistance gene-free complementation plasmid

If all antibiotic-resistance genes from the complementation plasmid were completely removed, selection could be made exclusively on auxotrophy-complementation as it has been reported, for example, by Dong *et al.*, (2010). Towards this objective, inverse PCR was performed on the plasmid pFC2 using primers 24 and 25 (Table 4.3) to amplify the complete plasmid while excluding the ampicillin and kanamycin resistance genes (Binnewitt, 2013). The theoretical reduced plasmid resulting from this modification is shown in Fig. 5.24.

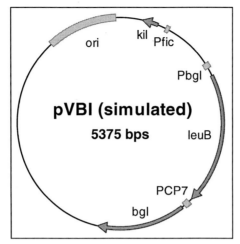

Fig. 5.24: Plasmid map of the hypothetical *leuB* complementation construct completely free of antibiotic resistance genes.

After a systematic series of optimizations including touchdown PCR, the generation of the linear PCR product was successful. This PCR product was subsequently verified by sequencing near the ends. After phosphorylation at the 5' ends using T4 polynucleotide kinase, it had to be ligated and recircularized before being transformed into chemical competent cells of the auxotrophic strain *E. coli* JW5807-2. The ligation reaction was confirmed to have been successful by analysing the ligated product on an agarose gel and verifying the difference in band size to the linear PCR product. The transformed strains had to selected on SGA medium since there was no possibility for antibiotic-based selection anymore in the reduced plasmid pVBI. Such a selection was also reported earlier in the work involving glycine auxotrophic mutants (Vidal *et al.*, 2008). According to transformation tests with the original plasmid pFC2 and subsequent selection on SGA medium, it was found that after the transformation, the cells had to be incubated in LB medium for 1 hour to allow for the *leuB* gene to be expressed followed by selection for positive transformants in shaking flasks with SGA medium. However, even after repeated attempts, this step proved to be challenging and, therefore, the selection of clones of pVBI based on leucine auxotrophy complementation was not successful.

5.2.9 Two-stage chemostat cultivation of *E. coli* JW5807-2-pFC2

While optimizing conditions for biomass growth, the ultimate aim was to achieve reasonable levels of the recombinant product in the extracellular medium. In order to check if a cascade of vessels would improve the productivity of the *E. coli* JW5807-2-pFC2 system for the target enzyme, a two-stage chemostat consisting of 2 reactors placed in series, (see Fig. 2.21) was setup. Such a system was expected to bring about a favourable age-distribution of cells where the first vessel receives fresh feed and would mostly be populated by young actively growing cells which when reaching the second vessel would have limited scope for further increase in numbers. Such 'old' cells are expected to show an activated stationary-phase response which includes activation of the P_{fic} promoter leading to a better excretion of product into the medium. Such a segregation of growth and product excretion would not be possible in a single vessel of comparable volume due to the homogenous age distribution of the cells. The flow rate through the reactors remained the same for both. While fresh medium entered the first reactor R1, the used medium, cells and products stream entered the second reactor R2. All process conditions were kept similar in both vessels and no antibiotics were added. The glycerol (limiting substrate) concentration in the feed was 20 g L^{-1}, each vessel had a working volume of 1 L and space velocities from 0.15 h^{-1} to 0.35 h^{-1} were tested.

The first stage would behave like a single-stage reactor that was used in all previous experiments. But the samples taken from the second stage are expected to provide new data. This is because, along with the second reactor which is simply connected in series to the first and therefore can be thought of as an extension, the overall apparent volume of the system has been doubled to 2 V (2 L). So, with the same uniform flow rate F, if the space velocity achieved in the first reactor is taken as D, then the effective space velocity of the overall system as seen from the sample from the second stage, should be lower. This should cause a substrate limitation in R2 while still maintaining higher flow rates and could bring about an improvement in the process productivity, since such substrate limited conditions should trigger more export of product into extracellular space. Apart from this, the setup can be used to generally study about biomass productivity relationships in a 2-stage cascade. At high dilution rates, more residual substrate is expected to be carried over from the first on to the second stage (Bakker *et al.*, 1996) and could therefore yield a better conversion while maintaining high flow rates.

To demonstrate the interplay between the two reactors in series, the following are values of dissolved oxygen saturation (DO) and CO_2 in the exhaust measured at two different steady states.

At $D = 0.1$ h^{-1}: R1: $DO = 10\%$; $CO_2 = 1.21\%$ R2 : $DO = 90\%$; $CO_2 = 0.12\%$

At $D = 0.2$ h^{-1}: R1: $DO = 20\%$; $CO_2 = 0.9\%$ R2 : $DO = 65\%$; $CO_2 = 0.6\%$

In the first case, the low feed rate gave enough residence time for the cells in the first reactor to consume more substrate and hence the reduced oxygen levels and high CO_2 levels. The second reactor showed very little scope for any further growth. In the second case, $D = 0.2$ h^{-1}, the higher feed rate decreased the residence time in the first reactor, hence increased the oxygen saturation levels, caused a reduction in biomass density and CO_2 levels in exhaust. However, the unused substrate flowing into the second stage gave more scope for metabolism and thus a decrease in oxygen levels and increase in CO_2 level was detected as compared to the lower feed rate.

Different feed rates (0.1 L h^{-1} to 0.35 L h^{-1}) were tested and the results obtained are presented in the following diagrams (Fig. 5.25). The samples measured from the second vessel (R2) depict the overall output from the system, which is twice the volume of the first stage. Hence, the space velocity gets halved in the case of the second reactor for the comparison with the case of a single stirred tank of equal total volume. Therefore, the volume for R2 is taken as 2 *V* which effectively halves the space velocities for the whole cascade. However, since the same flow rate is used, and the volumes of R1 and R2 are identical, when R1 nears washout point, R2 would be driven towards washout anyway.

A

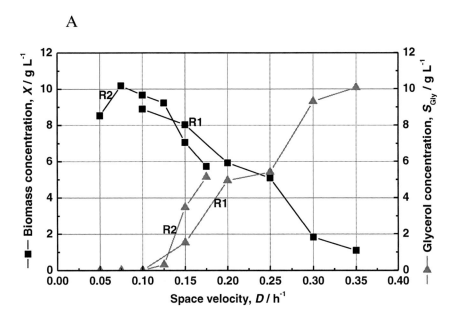

B

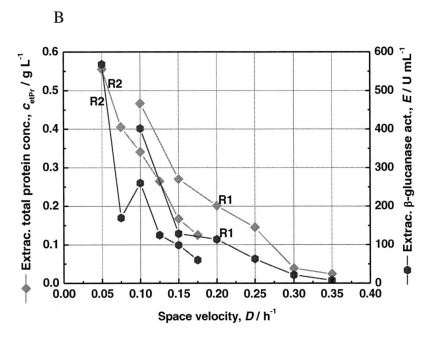

C

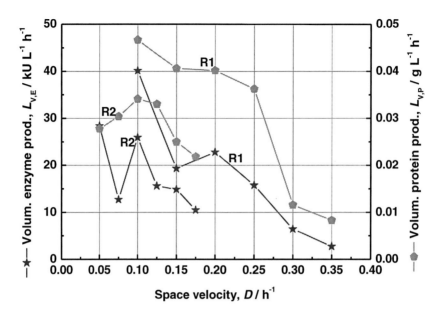

Fig. 5.25: Summary of results from the 2-stage chemostat. The values for R2 which is the second stage, represent the end values from the overall system and hence have been considered with a volume of 2 V, where V is the volume of any stage. (A) Biomass and substrate concentration profiles, (B) β-glucanase and total protein concentrations achieved in the extracellular fraction, and (C) volumetric productivities for β-glucanase and total protein.

In Fig. 5.25A, it is found that the second stage was able to achieve slightly higher biomass densities than the first stage, which is very much expected. It is also possible to find the similarity in this graph and that to the general schematic for 1st order kinetics presented below in Fig. 5.26. The biomass was sustained at 5.7 g L^{-1} in R2 during the point where there was almost a washout to be expected in R1. At a point just thereafter, R1 may have washed out followed closely by R2.

Fig. 5.25B shows the concentration profiles of recombinant β-glucanase and total protein measured in the extracellular fractions (cell-free supernatants). At every data point the absolute value of enzyme activity achieved from the second stage was greater than that from the first stage. This effect was more pronounced in the values from total protein concentration (see chapter 4.7.4). However, as it would be discussed below, this increase was not sufficient to translate into a gain in terms of productivity. The selectivity for the target protein at the lowest space velocity was 0.14 g g^{-1} and 0.2 g g^{-1} for R1 and R2 respectively. With increase of the space velocities, the selectivities in both the reactors declined uniformly to reach values of 0.02 g g^{-1} and 0.03 g g^{-1} at the condition of highest space velocity, and therefore no major increase could be achieved in the second vessel.

Fig. 5.25C representing the volumetric productivity, showed a distinctly contrasting profile. The volumetric enzyme productivity in R2 was initially lower, but later with higher relative space velocities, it remained stable at a higher value (15 kU L^{-1} h^{-1}) while the corresponding values from R1 decreased continually. Thus the recombinant product could be expressed at slightly higher volumetric productivities than a single vessel, when using near maximum space velocities. However, the 2-stage process ultimately did not bring any improvement in maximum volumetric productivities in comparison to a single vessel (see Fig. 5.21). An analysis of the kinetics behind multiple vessels in series, presented in the following section throws more light into why this was the case.

Analysis of multiple stage chemostat mass balance

Segregation of growth and production stages is a fairly well-known strategy for continuous cultivation and recombinant protein expression (Park *et al.*, 1990). However, in this work, the implementation of a 2-stage chemostat cascade did not result in the desired improvement of volumetric enzyme productivity. The following diagram (Fig. 5.26) shows the profiles for biomass, substrate concentrations and volumetric biomass productivity for a single or multiple stage chemostat process. Briefly put, the biomass sustains itself over a wider range of relative dilution rates inside the vessel when only one stage is used (blue). But it is also to be noted that it decreases over a longer period and already starts its downfall during lower relative space velocities. The introduction of a second stage (green) splits the volume and drastically reduces the maximum limit on relative dilution rate before wash out occurs. But, the biomass decrease is now far sharper and occurs very close to the wash out point and remains high for most of the time. This is because, the inlet stream entering the second stage is already populated with actively growing cells and it is not really a 'dilution' in the true sense. But at the point the first stage washes out, it immediately becomes equivalent to the feed inlet vessel and the second stage becomes the 'first' and only stage. Thus when feeding at the same rate, this stage should also wash out immediately thereafter. The trend continues in the same direction for any number of subsequent stages and it can be seen that the critical relative dilution rate moves in a periodic fashion (1/2, 1/3, 1/4... and so on). Due to the splitting up of the volume, the apparent critical dilution rate reduces uniformly in each stage as seen from the wash out points, although they show progressively better substrate utilizations at low D_{rel}. The exact inverse observation can be made for the residual substrate concentration according to equations (30) and (31).

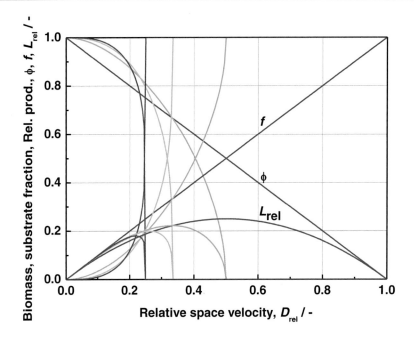

Fig. 5.26: Representation of the behavior of a single stage (blue) in terms of biomass (Φ) and residual substrate (*f*) fractions and also the relative biomass productivity (L_{rel}) under different relative dilution rates for first order kinetics. The same parameters show strikingly different profiles in the second (green), third (cyan) and fourth (red) stages connected in series. Maintenance energy is ignored. Source: Flaschel, Bielefeld University.

The volumetric productivity curves, clarify the problem as to what governs the final biomass output levels that can be achieved from this 2-stage cascade. The maximum biomass productivities in Fig. 5.26, never increase above the level achieved in the first stage. With regard to the volumetric productivity of the target enzyme, it is important to clarify the terms used for the calculation, especially in the case for a cascade. According to Davis *et al.*, (1990), when using multiple vessels,

$$\text{volumetric productivity} = \frac{\text{product concentration} \cdot \text{flow rate}}{\textbf{total} \text{ volume}}$$

which is simply product concentration multiplied by dilution rate for a 1-stage process. For the 2-stage process therefore, the final volumetric productivity calculation should take into account the volume of both the vessels, hence, 2 *V* or 2 L (here). This would require an enormous increase in actual product levels in the second reactor in order to translate into a corresponding increase in productivity. In the case of specific productivity, the same rule of total volume applies so that,

$$\text{Specific productivity } (L_{X,E}) = (E/X) \cdot F/V$$

Where *E* is the activity measured at the last stage and *X* is the corresponding biomass concentration at that stage, *F* is the volumetric feed rate and *V* represents the **total** volume of the system.

Nevertheless, the absolute volumetric enzyme activity achieved at every sample in R2 was higher than in R1 (Fig. 5.25B) which could be beneficial when subsequent purification steps are to be considered. To sum up, it could be said that the interplay between the growth and flow rates through the vessels has a large influence on the final output from the overall system (Fig. 5.26). In a continuous microbial process approaching exponential growth, more back-mixing would be anyway desired. But a multiple reactor system, connected in series theoretically approaches plug-flow behaviour. Although, the increased processing achieved in the higher stages resulted in better metabolism, the mass transport factor weighed in and countered any benefits accrued. This was seen from the fact that although R2 was fed with cells, any increase beyond the critical dilution rate of R1 would invariably have an effect on R2 also since the flow rate is the same. In other words, the advantage of a better utilization of substrate and sharper fall of biomass near the D_c is neutralized by the simple fact that the D_c occurs much earlier. So on the cellular level it means better substrate limited conditions but on the process side, there are constrictions due to the space velocity.

It is possible that if the chemostat experiment could have been continued for a couple of more increments in the feeding rate, eventually the wash out point of R2 (or the whole system) could have been determined. As shown for multiple-stage chemostats (Fig. 5.26), the curves for R2 occur over only a fraction of the whole plot area and reach their critical points closely after the first stage. A supplemental feed stream only to R2 containing concentrated medium and feeding at a very low flow rate could be attempted to ensure no large maintenance effects are created in the 2nd stage and the limiting substrate continues to be the same as that for the 1st stage (Fu et al., 1993). Another option would be to optimize the volume allocation between the two reactors that would maximize the volumetric enzyme productivity from the second reactor (Davis et al., 1990).

5.2.10 Characterization of strain E. coli JW5807-2 by growth in the presence of leucine

The Keio $\Delta leuB$ knockout strain E. coli JW5807-2 had proved to be an excellent host strain for complementation with plasmid pFC2 which enabled antibiotic-free stable cultivation and recombinant protein expression (see Figs. 5.19 and 5.21). However, the volumetric productivities achieved with this strain were consistently lower than those with the control strain E. coli JM109-p582. Therefore, it was required to study the growth of this knockout strain without the plasmid in batch and chemostat modes to derive basic information about its growth characteristics. To this end, a batch fermentation in 1 L volume was carried out. A process temperature of 30 °C was maintained to reflect the temperature used for the recombinant protein expression with pFC2. Since the strain was auxotrophic, the SGA medium was supplemented with 50 mg L^{-1} leucine. This concentration had already been tested under shake-flask conditions to be sufficient to overcome the auxotrophy (see Fig. 5.14). Since the Keio strains carry a resistance gene in place of the knocked out gene, kanamycin could be added at a final concentration of 50 mg L^{-1} to prevent any contamination.

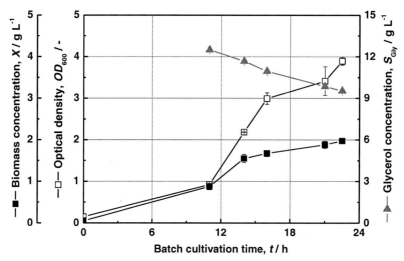

Fig. 5.27: Characterization of growth of strain *E. coli* JW5807-2 in SGA medium supplemented with 50 mg L^{-1} leucine by a batch fermentation process. The fermentation was carried out in the 2 L in-house fermenter according to the descriptions in Chapter 4.6.2.

The result from the batch fermentation (Fig. 5.27) however showed that growth of the strain stopped prematurely and that probably some other factor was limiting the growth of the strain. The biomass concentration only reached up to 2 g L^{-1} whereas there was still a residual glycerol concentration of 9.5 g L^{-1} present in the medium. Therefore, glycerol was not the limiting substrate as it is supposed to be but rather some other factor. The transport of leucine into the cells could have been a limiting factor.

In a separate trial, the strain was grown under similar conditions and tested in a chemostat at various space velocities to get more detailed information about its growth behaviour. Similarly, the batch fermentation was marked by a premature abortion of increase in optical density of the culture. The chemostat was nevertheless started and the strain allowed to stabilize at each of the space velocities. The feed medium contained leucine at a concentration of 50 mg L^{-1} and no antibiotic was included. The data from this chemostat experiment are shown in Fig. 5.28, and clearly proved the presence of some unidentified limiting factor that hindered the study of the growth kinetics. Another unexplained observation was the very low ratio of optical density to biomass concentration.

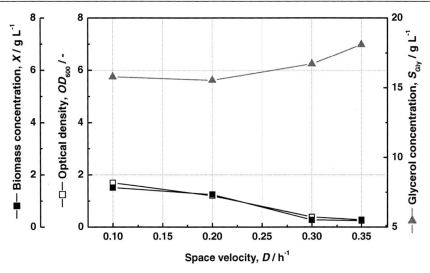

Fig. 5.28: Characterization of growth of strain *E. coli* JW5807-2 in SGA medium supplemented with 50 mg L^{-1} leucine by chemostat cultivation and analysis of steady states at various space velocities. The continuous cultivation was carried out in the 2 L in-house fermenter described in chapter 4.6.2 General experimental conditions for continuous culture were as described under chapter 4.6.3.

5.3 Auxotrophic system based on glycerol metabolism

The advent of biodiesel production has resulted in the mass production of crude glycerol as a by-product with very few identified applications (Hu & Wood, 2010). The use of glycerol in bioprocesses is therefore interesting and when used as sole carbon and energy source in a defined medium, the enzyme triose phosphate isomerase (TpiA) becomes immensely crucial since it links the glycerol metabolism to the glycolytic pathway. A blocking of this step would disable the cell to efficiently channelize glycerol from the medium into glycolysis but only accumulate it as methylglyoxal. Therefore, the second major strategy tested for antibiotic-free plasmid selection was the catabolic route of glycerol utilization and its connection to glycolysis. The results pertaining to this strategy starting from the construction of the genetic construct for complementation, its application in fermentation at various modes and finally reverting back to molecular genetic tools for improvement of this system would be presented in the following sections.

5.3.1 Cloning of *tpiA* gene expression cassette onto plasmid p582

To begin with, the major genes involved in glycerol transport and utilization (see Fig. 2.10) were tested to screen for a candidate auxotrophic gene target. Four knockout strains each with a deletion of one of the genes *tpiA*, *glpK*, *glpF* and *gldA* were screened to find out if they were susceptible to selection pressure when grown on minimal medium with glycerol as sole carbon and energy source. From these initial experiments it was found that the *tpiA* gene knockout strain could be a good candidate as an auxotrophic host strain (Velur Selvamani *et al.*, 2014). The knockout of the gene *tpiA* (coding for triose phosphate isomerase) blocked the conversion of dihydroxyacetone phosphate into glyceraldehyde-3-phosphate and hence halted the metabolic flow from glycerol utilization. When glycerol was used as the sole carbon source in the minimal medium, this could create a selection pressure for cells carrying a plasmid with a cloned copy of the *tpiA* gene for complementation.

The gene *tpiA* along with its preceding natural promoter (P1 and P2) sequences and end terminator sequences (total length 1089 bp) was amplified from *E. coli* K12 MG1655 genomic DNA using primers 5 and 6 (Table 4.3) as forward and reverse primers respectively. The amplified region started 150 bp upstream of the ATG start codon and ended 172 bp downstream of the TAA stop codon of the *tpiA* structural gene. The upstream region started in the *yiiQ* gene, coding for an unknown conserved protein, including the predicted *tpiA* promoter (Mendoza-Vargas *et al.*, 2009). The downstream region reached into the *cdh* gene including the predicted Rho-independent *tpiA* terminator (see Fig. 5.30; Keseler *et al.*, 2011). The blunt fragment was cloned into plasmid pJET (Thermo Scientific, Germany) and the sequence was verified. This construct called pJET-tpiA was used as a template for amplification of the *tpiA* fragment with EcoRI ends by PCR using primers 7 and 8 (Table 4.3). A gradient PCR with 4 different annealing temperatures was tested. Although two other regions were amplified and seen as light bands above and below the expected 1.1 kb region, the samples were pooled together, purified using a PCR cleanup kit (Promega GmbH, Germany), run again on agarose gel, and the desired band cut out and eluted from the gel. This DNA was used for EcoRI restriction to generate the insert. The restricted sample was purified by simply mixing with equal volume binding buffer and loading onto a silica column (both Promega GmbH, Germany). Subsequent ethanol washing and elution with autoclaved MilliQ water gave the final insert. The vector p582 was cut at its unique EcoRI site and subjected to phosphatase treatment to remove the 5'-phosphate groups. This vector-insert combination was used for the subsequent cloning steps. After successful cloning, 4 potential colonies were screened by restriction digestion (Fig. 5.29).

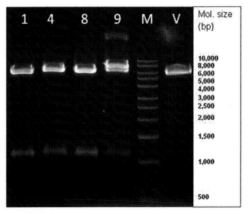

Fig. 5.29: EcoRI restriction and screening for positive recombinant clones. Numbers 1, 4, 8, and 9 refer to clones called pFC1, pFC4, pFC8 and pFC9. V refers to the restricted vector p582 as control. M refers to 1 kb molecular size marker from Plasmid Factory GmbH, Germany.

Since the natural promoter and terminator sequences were used, the cloned *tpiA* fragment contained the downstream portion of the preceding gene *yiiQ* included as part of the promoter sequences and the downstream portion of the succeeding gene *cdh* (*cdh* is in opposite orientation) included as part of the terminator sequences (Fig. 5.30). This arrangement still involved the risk that a recombination between the knockout location on the chromosome and the *tpiA* region on a plasmid could occur.

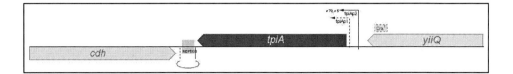

Fig. 5.30: Genetic organization of the gene *tpiA* in relation to its flanking genes. Adapted from Ecocyc.org (Keseler *et al.*, 2011).

Two constructs were verified by sequencing and named pFC1 and pFC4. Partial sequencing of pFC1 and pFC4 revealed that the regions of the cloned *tpiA* gene were oriented in opposite directions (Fig. 5.31).

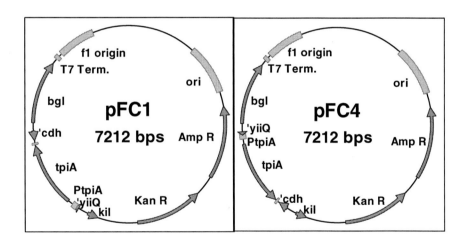

Fig. 5.31. Structure of plasmid constructs pFC1 and pFC4. The auxotrophy-complementing *tpiA* gene along with its natural promoter (PtpiA) and terminator sequences has been cloned in the forward orientation in pFC1 and in reverse orientation in pFC4.

5.3.2 Complementation of *E. coli* JW3890-2 with construct pFC1 and pFC4

The clones pFC1 and pFC4 were checked for their efficiency to complement the auxotrophy by transforming into competent cells of the Keio *tpiA* knockout strain *E. coli* JW3890-2 and cultivating in shake-flasks (Fig. 5.32). The strain JW3890-2 did not grow in SGA medium with glycerol as the only carbon source (SGA-Kan). However, if a complex carbon source was present (LB-Kan), some growth was observed. Only the strains complemented with plasmids pFC1 (JW3890-2-pFC1) or pFC4

(JW3890-2-pFC4) showed growth in SGA containing glycerol as the only carbon source (SGA-Kan). The strain JW3890-2-pFC1 showed a shorter lag phase than the strain with construct pFC4. In fact, this phenomenon could be reproduced over multiple trials. Both strains reached equal maximum optical densities (OD_{600}) of about 8.0. The media had been supplemented with kanamycin in all three cases.

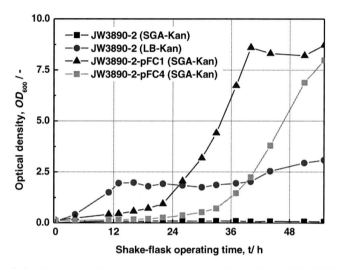

Fig. 5.32: Shake-flask cultivation to demonstrate auxotrophy complementation of *E. coli* JW3890-2 through plasmids pFC1 and pFC4. An incubation temperature of 30 °C was maintained with other general shake-flask cultivation conditions as per the descriptions in chapter 4.6.1.

When using the Keio knockout strain *E. coli* JW3890-2, a flux from DHAP into the xylulose pathway is not possible due to the deletion of rhamnose metabolism genes (see 4.1.2; Baba *et al.*, 2006). A bypass of DHAP to lactate and pyruvate through methylglyoxal is possible through the glyoxalase I-II pathway (Ferguson *et al.*, 1998). This was probably the reason for the limited growth of the cells without the complementation plasmid in complex medium. However, when glycerol is the sole carbon and energy source, the toxicity of methylglyoxal should render this route unfavourable. Moreover, this route only represents a possible metabolic by-pass and not an energy-efficient pathway to give a growth rate advantage. Although, both clones were able to complement the auxotrophy of the *tpiA* Keio knockout strain, their product expression capabilities differed considerably. Both constructs were studied under batch cultivation conditions in a bioreactor.

5.3.3 Batch fermentations with strain *E. coli* JW3890-2- pFC1

The strain *E. coli* JW3890-2-pFC1 was cultivated in the mini-fermenter with 1 L working volume. SGA medium with glycerol as sole carbon source was used without the addition of any antibiotics. As seen with the shake-flask cultivation, the batch process also showed that the plasmid pFC1 could complement the *tpiA* gene knockout and enable the strain to grow with glycerol as the sole carbon source (Fig. 5.33). A biomass concentration of 8 g L^{-1} was reached at an operating time of 23 h. The accumulation of extracellular proteins was delayed reaching a maximum of 0.4 g L^{-1}. The extracellular concentration of β-glucanase accumulated in parallel to the profile of total extracellular protein, but reached a volumetric activity of only 68.4 U mL^{-1}. The ratio of the enzyme activity to the extracellular total protein concentration (see chapter 5.2.4) was as low as 0.19 kU mg^{-1} and the reason for this observation could not be ascertained. To note again, this was the construct in which the *tpiA* gene had been cloned in the forward orientation with respect to the *kil* and *bgl* genes on the plasmid p582.

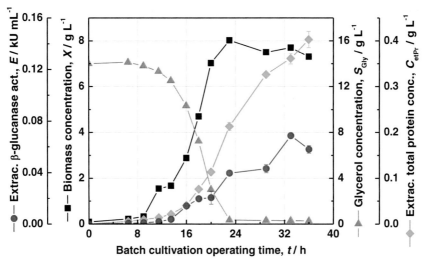

Fig. 5.33: Profiles of fermentation products during batch fermentation with *E. coli* JW3890-2-pFC1. The fermentation was carried out at a temperature of 30 °C in the in-house fermenter according to the descriptions in chapter 4.6.2.

The batch process was also carried out in a 7 L NLF 22 fermenter (Bioengineering AG, Switzerland) with a working volume of 5 L (see chapter 4.6.2). This trial resulted in a very low and almost negligible extracellular β-glucanase activity. The maximum activity level reached at the end of 41 h was 13.7 U mL^{-1}, which was well into the stationary phase with an optical density of 19.9. Therefore, the enzyme activities in the cytoplasmic and periplasmic fractions were assayed and they were quite higher at 59.9 U mL^{-1} and 43.2 U mL^{-1} respectively (maximum levels). The samples were visualized on an SDS-PAGE gel shown in Figure 5.34. Since the enzyme was highly diluted in the extracellular samples, it had to be concentrated by TCA precipitation before loading (see chapter 4.7.8).

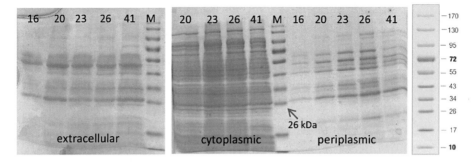

Fig. 5.34: (left) Extracellular fractions from different time points shown in the lanes, were loaded onto 12% SDS gel after TCA precipitation. (right) cytoplasmic and periplasmic fractions of the same samples loaded directly without any concentration. The supposed recombinant beta-glucanase protein is expected at the 26 kDa position which is shown here by an arrow and is the 4th band below the orange-dyed 72 kDa band in the standard marker (M). Molecular mass standard SM0671 (Fermentas, Lithuania) is shown separately with molecular masses given in kDa.

5.3.4 Batch fermentations with strain *E. coli* JW3890-2- pFC4

From the batch fermentation with strain *E. coli* JW3890-2-pFC4 (Fig. 5.35), it was immediately apparent that this construct was able to achieve higher extracellular recombinant enzyme activities as compared to the strain JW3890-2-pFC1. In fact, the maximum extracellular β-glucanase activity reached 132 U mL^{-1} at an extracellular total protein concentration of 0.26 g L^{-1}. A maximum biomass concentration of 6.35 g L^{-1} was achieved after 24 h of process time.

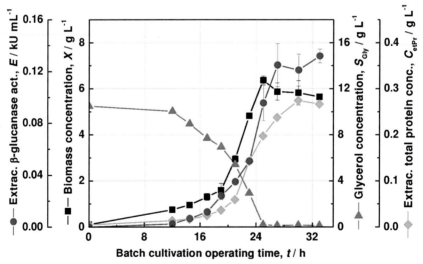

Fig. 5.35: Profiles of fermentation products during batch fermentation with *E. coli* JW3890-2-pFC4. The fermentation was carried out at a temperature of 30 °C in the in-house fermenter according to the descriptions in chapter 4.6.2.

However, the ratio between extracellular enzyme activity and extracellular total protein concentration was reduced to about 0.5 kU mg^{-1} compared to the higher values for the reference and pFC2 systems seen in chapter 5.2.4. This suggested some kind of limitation in export of the target enzyme across the outer membrane. Similarly, the selectivity of target recombinant product with respect to biomass ($S_{P/X}$) was a low 0.06 g g^{-1} in comparison to the selectivities of the reference or the pFC2 system (see chapter 5.2.4).

Similar to the previous strain, a batch fermentation was also carried out with the strain JW3890-2-pFC4 in a NLF 22 fermenter with a working volume of 5 L. The fermentation was operated for a total period of 48 h. Two major inferences could be made to the fermentation with construct pFC1. The maximum extracellular recombinant enzyme activity stood at 83.4 U mL^{-1}. Intriguingly, just as in the case of pFC1, the periplasmic fraction for the fermentation with pFC4 also showed a higher activity (163.1 U mL^{-1}) than the extracellular fraction. With the reference strain *E. coli* JM109-p582, the periplasmic activity had (for low dilution rate values) remained at a level lower than the extracellular activity. In other words, the distribution was towards excretion of product into the extracellular space. In the present case, it seemed, the *kil* gene expression, even at stationary phase conditions, was somehow not optimally achieved, so that a higher activity of the recombinant enzyme remained within the periplasm. The following image (Fig. 5.36) shows the extracellular samples from different time points before and after TCA precipitation.

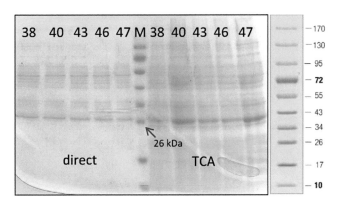

Fig. 5.36: Extracellular fractions corresponding to different time points from batch fermentation with strain *E. coli* JW3890-2-pFC4, loaded onto 12% SDS gel directly (lanes 1-5) and after TCA precipitation (lanes 7-11). M refers to molecular mass standard SM0671 (Fermentas, Lithuania) which is also shown separately.

It could be observed over multiple trials that the strain JW3890-2-pFC1 grew with a slightly shorter lag phase and achieved higher maximum biomass concentrations than JW3890-2-pFC4. Again in multiple trials, with respect to the extracellular recombinant β-glucanase activity, the construct pFC4 clearly returned higher volumetric activity than pFC1. With regard to this difference in product expression capability, one aspect to consider would be that, although the sequenced regions covering the cloned *tpiA* gene were found to be in order, the other regions of the plasmid could have undergone a mutation at some critical position since the insert sizes were slightly different between pFC1 and pFC4 (Figure 5.29). With respect to secretion of product into extracellular medium, both the clones pFC1 and pFC4 were found to be poor. In general, the reference strain *E. coli* JM109-p582 showed up to 10 times the activity achieved with the *tpiA* auxotrophic strains.

5.3.5 Antibiotic-free continuous cultivation of *E. coli* JW3890-2-pFC4

Since the clone pFC4 yielded a higher activity than pFC1 in batch cultivations, it was chosen for the chemostat studies. A chemostat was set up in the minifermenter with a 2 L total volume in which a working volume of 1 L was maintained. The SGA minimal medium based on glycerol as the only carbon source was used. The feed solution contained 20 g L^{-1} glycerol. No antibiotics were added to the media. At a space velocity of 0.1 h^{-1} and maintaining antibiotic-free conditions, the biomass concentration was stabilized at about 10.5 g L^{-1}. A temporary loss of steady-state conditions at 125 h resulted in disturbances over all the measured parameters, but it could be seen that eventually the curves show a tendency to recover back to their original levels. A steady state extracellular recombinant β-glucanase activity of 0.1 kU mL^{-1} could be maintained under antibiotic-free conditions.

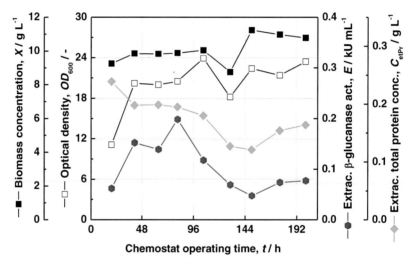

Fig. 5.37: Chemostat cultivation of *E. coli* JW3890-2-pFC4 at D = 0.1 h^{-1}. The cultivation was carried out under antibiotic-free conditions at a process temperature of 30 °C in the in-house fermenter described in chapter 4.6.2. Other general cultivation conditions were as described in chapter 4.6.3.

The plasmid segregational stability was maintained based on the auxotrophy complementation principle for over 200 h (Refer to Fig. 5.62). The stability of the isolated plasmid concentration normalized to optical density for the strain *E. coli* JW3890-2-pFC4 was followed and compared to the reference strain *E. coli* JM109-p582 which was shown earlier in Fig. 5.8. Whereas for the reference strain, a constant space velocity of 0.1 h^{-1} had been maintained, the space velocity for the complementation strain was varied with a profile of 0.2 h^{-1} for 75 h, 0.35 h^{-1} for 21 h, 0.3 h^{-1} for 19 h, 0.1 h^{-1} for 77 h, 0.25 h^{-1} for 17 h, and 0.15 h^{-1} for of 26 h. As seen before, in the absence of antibiotics, the strain JM109-p582 lost its plasmid continually, whereas the plasmid concentration for the strain JW3890-2-pFC4 stabilized at an average value of 10.6 ng μL^{-1} (Fig. 5.38).

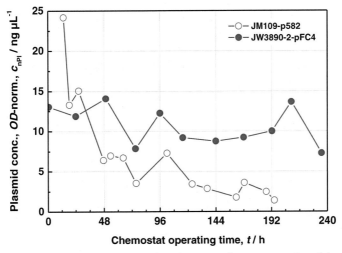

Fig. 5.38: Comparison of plasmid content of cells from long-term chemostat operation of the reference strain JM109-p582 and *tpiA* auxotrophic complementation system JW3890-2-pFC4.

In separate growth measurements in shake-flasks, the complemented strain JW3890-2-pFC4 showed a maximal specific growth rate of $0.265\ h^{-1}$. At the end of 96 h and 192 h, 10 swabs from the antibiotic plate used for the replica plating test, were grown overnight, and their plasmids were analysed after EcoRI restriction (Figure 5.39). The expected 1.1 kb insert corresponding to the *tpiA* gene could be verified. Only 1 colony showed an inconsistent band probably due to a restriction reaction inconsistency (lane 1 from 192 h).

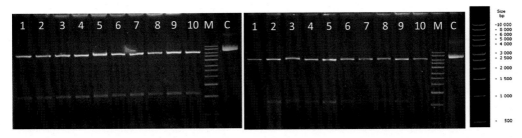

Fig. 5.39: Ten colonies randomly chosen at 96 h (left) and 192 h (right) of chemostat operation with *E. coli* JW3890-2-pFC4. The plasmids were analysed after EcoRI restriction. Last lane (C) shows reference sample of one of the unrestricted preparations as a control. M is 1 kb molecular size ladder (Plasmid Factory GmbH, Germany) also shown separately for comparison.

5.3.6 Testing different space velocities for continuous cultivation of *E. coli* JW3890-2-pFC4

The chemostat experiments carried out at different space velocities offered further information about the kinetic relationships involved between growth and product expression in the strain *E.coli* JW3890-2-pFC4. No antibiotics were used at any point during the process. The declines of the cell density as well as the extracellular enzyme activity were linear functions of the space velocity (Fig. 5.40). The decline of cell density started from about $0.15\ h^{-1}$. The extracellular enzyme activity started declining already from 0.1 h⁻¹. The maximum volumetric enzyme productivity was even lower than the *leuB* complementation system and stood at about 6.55 kU L⁻¹ h⁻¹. However, the plasmid stability by replica plating was almost absolute throughout the whole process.

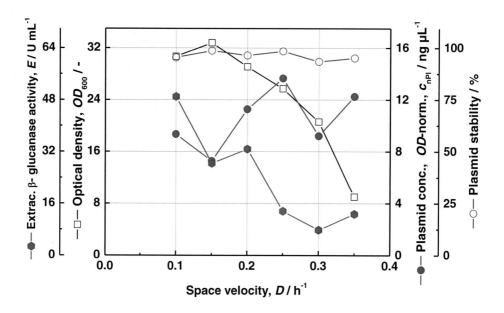

Fig. 5.40: Chemostat with *E. coli* JW3890-2-pFC4 over different space velocities. The cultivation was carried out under antibiotic-free conditions at a process temperature of 30 °C in the in-house fermenter described in chapter 4.6.2. Other general cultivation conditions were as described in chapter 4.6.3.

Unlike in the case of the reference strain JM109-p582 (see Fig. 5.10) or the strain JW5807-2-pFC2 (see Fig. 5.21), no distinct relationship between isolated plasmid concentration and growth rate could be established for the strain JW3890-2-pFC4 from the profile in Fig. 5.40.

Thus, an alternative plasmid selection principle based on the complementation of *tpiA* for the continuous cultivation and recombinant expression of β-glucanase could be established. In comparison to anabolism-based systems, very few catabolism-based auxotrophy complementation systems for plasmid selection in bacteria have been described. In one of the earliest reports of a catabolism-based system, the essential *eda* gene coding for 2-keto-3-deoxy-6-phosphogluconate aldolase (KDPG-aldolase) in *Ralstonia eutropha* was cloned onto the recombinant plasmid for expression of cyanophycin synthetase and maintained with high stability in a *R. eutropha eda* deletion strain (Δ*eda*) when grown in minimal medium with gluconate or fructose as sole carbon and energy source. Since the targeted gene represented a key intermediate step in an exclusive conversion flux of these substrates towards glyceraldehyde-3-phosphate and pyruvate (Entner-Doudoroff pathway), a high selection pressure for maintenance of the plasmids was possible resulting in a loss of only 7% in the fraction of plasmid-bearing cells and a resulting high yield of cyanophycin in fed-batch fermentations at the 30 L and even 500 L scale (Voss & Steinbüchel, 2006). The attractiveness of the catabolism-based strategy described in the current work is that, although restricted to glycerol as a carbon source, it represents an interesting utilization avenue for this excessively produced by-product from the biodiesel industry. Moreover, some freedom in choice of carbon source should still be possible without loss of selection pressure, since the deletion of *tpiA* is a blockage of the central carbon metabolism leading ultimately to disadvantaged specific growth rates and toxicity due to the generation of methylglyoxal. This was particularly true for the use of glucose as sole carbon source in SGA medium which was found not to affect the selection pressure (Velur Selvamani *et al.*, 2014).

Although, multiple genomic integration of heterologous genes has resulted in the possibility to stably maintain the target gene in antibiotic-free conditions and also use complex media without danger of horizontal effects (Tyo *et al.*, 2009), the antibiotic resistance marker would still be present on the final strain. Moreover, the antibiotic resistance gene would be present in multiple copies which would add to unnecessary metabolic burden. On the other hand, in the system described in this work, the Keio strain always offers the possibility of removal of the kanamycin resistance cassette from the genome by the action of a temperature-sensitive curing plasmid such as pCP20 which confers the action of FLP recombinase (Datsenko & Wanner, 2000).

5.3.7 Modification of plasmid pFC4 by replacing the natural promoter PtpiA with artificial ones CP19 and CP33

The problems associated with over-expression of an auxotrophic complementation gene have been discussed in the study involving *glyA* knockout complementation by serine hydroxymethyl transferase (SHMT) and the idea of using a weak constitutive promoter to control expression of the complementation gene was invoked (Vidal *et al.*, 2008). Since a low level of expression of the *tpiA* gene from the plasmid pFC4 would be sufficient to complement the auxotrophy in the strain JW3890-2, the natural P1/P2 promoter system (see Fig. 5.30) of PtpiA was to be replaced with a weak promoter chosen from a library of synthetic promoters for *E. coli* (Jensen & Hammer, 1998). According to their study, the promoters CP19 and CP33 have strengths of 3.3 and 7 Miller units respectively in *E. coli*. This would modify the pFC4 construct and force the cell to maintain a higher content of the plasmid. The plasmid pFC4 has a pUC origin of replication and, therefore, should appear at high copy number. This points to a possibility of excessively high copies of *tpiA* and thereby unnecessary transcription and stress for the cells. The effect of having a promoter with moderate or low activity for *tpiA* on the plasmid was to be tested to find out if it would help in improving the overall efficiency of the host cells. The promoters CP19 and CP33 were both classified as relatively weak by the original study (Jensen & Hammer, 1998). Since the target plasmid pFC4 has a size of 7.2 kb, it was found to be easier to perform the promoter change on the original smaller pJET-tpiA instead, followed by cloning the *tpiA* gene including the new promoter back into the reference plasmid p582 in a step similar to the one described for the construction of pFC4 (see text in chapter 5.3.1 for Fig. 5.29). The sequences of the artificial promoters CP19 and CP33 are shown below starting from the -52 site. The -35 and -10 regions have been highlighted. The promoters were chosen only up to the transcription start point since the region starting from the transcription start site onwards was to be carried on the plasmid. The effects of changing exclusively the promoter sequence alone are to be studied.

CP19

CATCGCTT**AGTTTTTCTTGACA**GGA-GGGATCCGGG**TTGATATAATA**GTTA

CP33

CATGTTGG**AGTTTATTCTTGACA**TAC-AATTACTGCAGT**GATATAATA**GGTGA

The promoters were split into 2 halves and each half was added as a 5'-overhang to primers that bound and amplified the entire plasmid excluding the target region spanning the P1/P2 promoter. In the schematic diagram in Fig. 5.41, primer (19/33)-1 contained the second half of the new promoter as overhang while primer (19/33)-2 contained the reverse complement of the first half as overhang.

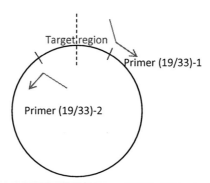

Fig. 5.41: Schematic representation of inverse PCR method for replacing a promoter sequence on a plasmid with a new sequence. The 5'-overhangs containing the synthetic promoter sequences are shown in green.

The target promoter on the plasmid pJET-tpiA to be replaced has the sequence (5'-3'):

GGTTTGAATAAATGACAAAAAGCAAAGCCT-TTGTGCCGATGAATCTCTATACTGTTTCAC

Primers containing the synthetic promoters as overhangs are given in Table 4.3. Since the long non-binding overhangs of the primers had completely different melting temperatures as the binding regions, the annealing characteristics would be very different between the beginning of the PCR and after the first few cycles.

Promoter CP33:

In the case of CP33, the following PCR reaction (Table 5.1) was setup with primers 33-1, 33-2 and a 2-part cycling protocol including 3 different annealing temperatures during the first part (Table 5.2) was programmed (Fischer, 2013).

Table 5.1: Reaction setup for inverse PCR for the generation of a linear amplicon for promoter modification of pJET-tpiA using CP33 promoter.

Component	Volume per reaction (μL)
Template pJET-tpiA (18 ng μL^{-1})	0.3
5x Phusion GC buffer	10
10 μM primer 33-1	5
10 μM primer 33-2	5
10 mM dNTP mix	1.5
Phusion High Fidelity polymerase 2 U μL^{-1}	0.75
Nuclease-free water	27.45
Total	**50**

Table 5.2: 2-part cycling with gradient annealing temperature in the first part for inverse PCR for modification of pJET-tpiA using the CP33 promoter.

Step		Temperature (°C)	Time (s)
Initial denaturation		98	30
Denaturation		98	10
Annealing	15x	64-65	20
Extension		72	85
Denaturation		98	10
Annealing + Extension	20x	72	105
Final extension		72	600

Upon checking on an agarose gel, all the 3 temperatures tested showed excellent amplification. The reactions were pooled together, purified using a PCR clean up kit (Promega GmbH, Germany) and loaded again onto an agarose gel in order to selectively cut out only the desired 4 kb linear amplicon. This excised DNA band was again purified from the agarose gel and concentrated by sodium acetate precipitation. This gave a linear DNA product with a concentration of 135.6 ng μL^{-1}, a A_{260}/A_{280} ratio of 1.87 and a A_{260}/A_{230} ratio of 1.97. The linear DNA was phosphorylated at the 5' ends using the enzyme T4 polynucleotide kinase and the T4 DNA ligase buffer (both Thermo Scientific, Germany), since the enzyme was 100 % active in this buffer and the following ligation reaction could be conveniently continued. Table 5.3 shows the scheme for the phosphorylation reaction.

Table 5.3: Reaction setup for phosphorylation of a linear amplicon after inverse PCR for promoter modification on pJET-tpiA with CP33.

Component	Volume per reaction (µL)
Linear DNA product (135.6 ng µL^{-1})	10
10x T4 DNA ligase buffer	2
10 mM ATP	2
T4 polynucleotide kinase 10 U µL^{-1}	1
Nuclease-free water	5
Total	**20**

Phosphorylation was carried out at 37 °C for 40 min followed by deactivating the enzyme by incubation at 75 °C for 10 min. For the circularization by self-ligation the following reaction scheme (Table 5.4) was applied.

Table 5.4: Reaction setup for self-ligation of phosphorylated linear DNA for the promoter modification on pJET-tpiA with CP33.

Component	Volume per reaction (µL)
Phosphorylation reaction	1
10x T4 DNA ligase buffer	5
50% Polyethylene glycol (PEG 4000)	5
T4 DNA ligase (1 U µL^{-1})	5
Nuclease-free water	34
Total	**50**

Ligation was carried out at 4 °C overnight. After transformation into chemical competent *E. coli* Top 10 cells, successful clones were backed up on LB-ampicillin (100 µg mL^{-1}) agar plates and subjected to plasmid isolation. These plasmids were first analysed by restriction testing using the enzyme BpiI (BbsI). This enzyme cut only positive clones twice since the second restriction site was created only through the presence of the new promoter. In Fig. 5.42, this analysis is shown. The clones on lanes 1 and 3 clearly show the released sequence of 400 bp and are potentially positive clones. The small size of the released band could be further compared to the 400 bp band from an additional low-range DNA ladder. To the right of the markers M_1 and M_2, all the 4 clones have been loaded without restriction.

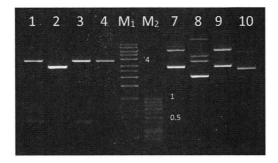

Fig. 5.42: Restriction digestion of potential clones for CP33 using the restriction enzyme BpiI. Lanes 1-4 show restricted DNA while lanes 7-10 show corresponding unrestricted DNA. M_1 is the 1kb DNA ladder from Plasmid Factory GmbH, Germany for which the 4 kb band position has been marked and M_2 is a100 bp DNA ladder from Fisher Scientific-Germany GmbH for which the positions of the1 kb and 0.5 kb bands have been marked.

The two positive clones (1 and 3) were sequenced for confirmation and stored as glycerol stocks with the designation pJET-CP33tpiA for the constructs. One probable cause for repeated failed attempts in this experiment seemed to be primer degeneration. If the primers were not HPLC-purified, there was a chance of the DNA suffering damages especially at the ends. Since the 5' ends contained the sequences for the new promoter which later needed to circularize, damages to the primer ends was more critical in this experiment than in a typical cloning procedure.

Promoter CP19:

The inverse PCR reaction for amplification of the construct with promoter CP19 using primers 19-1 and 19-2 (Table 4.3) was carried out similar to the reaction for CP33. However, Phusion HF buffer was used in place of GC buffer. The cycling conditions were also modified as shown in Table 5.5.

Table 5.5: Cycling conditions for inverse PCR for promoter modification with CP19 in pJET-tpiA.

Step		Temp (° C)	Time (s)
Initial denaturation		98	30
Denaturation		98	5
Annealing	35x	Gradient 60-70	20
Extension		72	85
Final extension		72	600

The gradient PCR resulted in a clear amplification of the required 4 kb amplicon as shown in Fig. 5.43. Since it was found that one of the forms of the template plasmid coincided in position on the gel with the expected 4 kb amplicon, it was required to carry out a DpnI digestion of the PCR reaction in order to disintegrate the template plasmid DNA since DpnI restricted only methylated DNA.

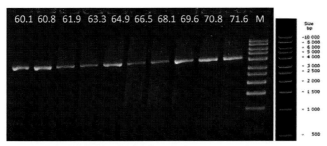

Fig. 5.43: Gradient PCR showing different annealing temperatures in °C for the generation of the amplicon with CP19 by inverse PCR. Standard marker M is 1 kb molecular size ladder from Plasmid Factory GmbH, Germany which is also shown separately for comparison.

DpnI digestion

A test of the activity of DpnI in various buffers revealed that the enzyme was able to restrict the plasmid pJET-tpiA equally effective in Phusion HF buffer as in the control Tango buffer (Fig. 5.44).

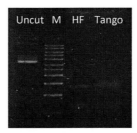

Fig. 5.44: Activity of DpnI in different buffers.

All the solutions from the PCR reactions shown in Fig. 5.43 were pooled together and treated with 1 µL of DpnI and incubated at 37 °C for 1 h. The enzyme was later deactivated by incubating the mixture at 80 °C for 20 min. The sample was purified, loaded onto 1% agarose gel and the target 4 kb amplicon was extracted. The extracted linear fragment was phosphorylated using T4 polynucleotide kinase according to guidelines by the manufacturer (Thermo Scientific, Germany) with respect to the amount of linear DNA ends (scheme shown in Table 5.6). The phosphorylated DNA fragment was self-ligated (recircularization scheme shown in Table 5.7) using T4 DNA ligase. Chemical competent cells of *E. coli* JM109 (Promega GmbH, Germany) were used for transformation of the ligated DNA.

Table 5.6: Reaction setup for the phosphorylation of linear PCR products after inverse PCR amplification for promoter modification on pJET-tpiA with CP19.

Component	Volume (µL)
Linear PCR product (125 ng µL^{-1})	10
10x reaction buffer A	2
10 mM ATP	2
T4 polynucleotide kinase	1
Water	5
Total	**20**

Table 5.7: Reaction setup for the ligation of phosphorylated linear DNA leading to recircularization for promoter modification on pJET-tpiA with CP19.

Component	Volume (µL)
Phosphorylated DNA fragment (62 ng µL^{-1})	1
10x T4 DNA ligase buffer	5
T4 DNA ligase (1 Weiss unit µL^{-1})	5
Water	39
Total	**50**

Screening of the transformants

The new promoter CP19 introduced a unique BamHI site which was not present in the original plasmid containing the natural *tpiA* promoter. Hence, restriction using BamHI could be used for screening potentially successful clones. After an initial round of testing, restriction was repeated on 3 potential colonies - 5, 19, 30 and compared with the template plasmid pJET-tpiA as negative control (Fig. 5.45). All three colonies displayed a clear restriction while in the case of the negative control, some changes were found but a single linearised band could not be found.

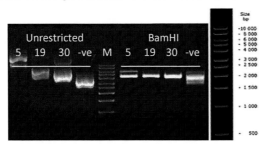

Fig. 5.45: Restriction digestion screening for transformants with a correct CP19 promoter sequence. Standard marker M is 1 kb molecular size ladder from Plasmid Factory GmbH, Germany which is also shown separately.

Upon sequencing the plasmid DNA from these colonies, it was found that colonies 19 and 30 contained base pair losses in the CP19 promoter ranging from 2 to 14. This could not have been due to the exonuclease activity of Phusion on the primers since the losses were all consistently on one side of the promoter only which corresponded to the 5' end of the primer. Phusion exhibits exonuclease activity from 3' to 5' direction only. These colonies showed up as positive during restriction screening because, the BamHI site was present completely on the second half of the new promoter, which was the half contributed by the other primer. Therefore, it was suspected that atleast in one of the primers, some base pair damages might have taken place in the 5' end. The colony 5 was verified for the presence of the correct CP19 sequence and the gene *tpiA*. This strain containing pJET-CP19tpiA was stored in the form of a glycerol culture.

These constructs with promoters CP19 and CP33 were transformed into strain JW3890-2 in order to test their ability to express the *tpiA* gene and complement the auxotrophy. The following diagram (Fig. 5.46) shows the results from the complementation test which was carried out in SGA medium in shake-flasks. Since baffle-less flasks were used, the growth was seen to occur after a long delay, or it may have been due to the weak-promoters. It was found that both the promoters could express the *tpiA* gene and result in growth of the auxotrophic strain in SGA medium in comparison to the strain without plasmid.

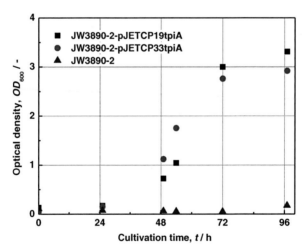

Fig. 5.46: Verification of auxotrophy complementation in *E. coli* JW3890-2 by expression of *tpiA* under the control of the weak promoters CP19 and CP33. The cultivations in SGA medium were carried out in baffle-less shake-flasks at a process temperature of 30 °C according to the general descriptions given in chapter 4.6.1.

The *tpiA* gene with modified weak promoters CP19 and CP33 were amplified from the constructs pJET-CP19tpiA and pJET-CP33tpiA respectively using primers 7 and 8 (Table 4.3). The inserts were prepared by EcoRI restriction as described earlier for the cloning of pFC4 and pFC1 and were cloned into the original vector p582 at its EcoRI site. Two colonies each for p582-CP19tpiA and p582-CP33tpiA were verified by restriction test as well as by sequencing. The sequencing results showed that all the plasmids contained the correct sequences of the new promoters and that the cloned fragments were all in the 'pFC4-type' orientation, i.e. the *tpiA* gene was present in the reverse orientation with respect to the *bgl* gene. All the constructs were transformed into chemical competent cells of the knockout host *E. coli* JW3890-2 and stored as glycerol cultures. These constructs were basically the same as the construct pFC4 shown in Fig. 5.31, the only difference being that the natural *tpiA* promoter P1/P2 had been replaced with CP19 or CP33. As an example, the sequence region for p582-CP33tpiA is shown in Fig. 8.5 in the Appendix.

Final testing of efficiency of expression

The strain *E. coli* JW3890-2 transformed with the constructs p582-CP19tpiA or p582-CP33tpiA was grown in SGA medium and compared to the strain transformed with the plasmid pFC4 in order to conclusively test the effect of the new promoters and check if they brought any improvement in the recombinant β-glucanase expression. The strains were cultivated in shake-flasks in SGA medium supplemented with kanamycin at a process temperature of 30 °C. The results shown in Fig. 5.47 suggested that the regulation of expression of the complementation gene *tpiA* by a weak promoter aided in improving biomass yields. The maximum optical densities achieved in the strains with the promoters CP19 or CP33 were markedly higher than that reached with the plasmid pFC4 containing the *tpiA* gene under its natural promoter. However, the better biomass growth failed to convert into a significant gain in recombinant product expression in all but one clone containing the CP33 promoter, in which a clear increase in extracellular recombinant β-glucanase activity could be observed at the onset of stationary phase. Therefore, this strain *E. coli* JW3890-2-p582-CP33tpiA clone 2 could serve as a useful starting point for future optimizations of the *tpiA* complementation strategy for antibiotic-free recombinant protein production.

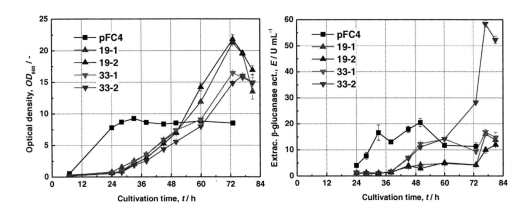

Fig. 5.47: Growth (left) and extracellular β-glucanase activity (right) from shake-flask cultivation for comparison of the efficiency of strain *E. coli* JW3890-2 complemented with plasmid pFC4 and 2 duplicate strains each for the modified plasmids containing promoters CP19 or 33 for the control of the *tpiA* complementation gene.

5.3.8 Creation of custom *tpiA* gene deletion in strains *E. coli* BW25113 and *E. coli* JM109

As mentioned under Theory, the Keio strains carried a deletion of the target gene according to a uniform fashion, in that only the portion between the start codon and the last 6 codons before the stop codon had been replaced with the kanamycin resistance gene. This presented a potential risk with the use of the strain *E. coli* JW3890-2 complemented with the construct pFC4 since the wild-type upstream, promoter, transcriptional start, coding sequence, terminator and downstream sequences had been used here which shared long homologous regions with the corresponding genomic regions present flanking the kanamycin resistance cassette. Any recombination between the plasmid and the genome by these sequences, could result in the loss of auxotrophy. Therefore, improved strains were generated that carried a wider deletion of the *tpiA* gene on the genome and thereby would not pose a risk of recombination.

Since the Keio knockout strain *E. coli* JW3890-2 was the strain tested earlier, logically, the Keio parent strain *E. coli* K12 BW25113 was chosen for a new *tpiA* knockout using custom-designed primers and encompassing a wider region upstream and downstream of the target. The primers Up and Low (Table 4.3) were designed to completely replace the *tpiA* gene along with the promoter (running into the upstream gene *yiiQ*) and the terminator region (running into the downstream gene *cdh*). This was exactly the region that was cloned into the plasmid pJET (pJET-tpiA) and used for testing complementation and is shown in Fig. 8.7 in the Appendix. A BLAST analysis of the homology arms of the selection cassette showed that they were highly specific in the genome (E value was 7×10^{-22} with the following match at E value of 2.0). To test the recombination in parallel in 2 strains, *E. coli* KRX and *E. coli* BW25113 were transformed with the plasmid pRed/ET which coded for the enzymes required for the recombination. The genomes of neither of these two strains were available, therefore the genome of *E. coli* K12 MG1655, which was closely related was used for *in silico* analysis with the program Clone Manager (all three are K strains and showed no differences in *tpiA*, *yiiQ* or *cdh*).

5.3.8.1 Gene deletion in BW25113

Generation of linear cassette

Using the primers Up, Low (Table 4.3), Phusion HF buffer and Phusion polymerase (4.5.3), a gradient PCR was performed for amplification of the selection cassette from the template plasmid pFRT (Table 4.2). All the tested temperatures showed amplification in the expected size (1.7 kb) as shown in Fig. 5.48. The PCR solutions were pooled together, electrophoresed again on an agarose gel and the target band was extracted.

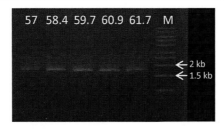

Fig. 5.48: Gradient PCR for the generation of a linear cassette for homologous recombination. The annealing temperatures in °C are shown above the lanes. M refers to 1 kb molecular size ladder from Plasmid Factory GmbH in which the bands 2 kb and 1.5 kb have been marked.

It was very important to carry out a DpnI digestion of the PCR solution in order to disintegrate the template plasmid pFRT and prevent it from entering the cells during electroporation. DpnI cuts only methylated DNA at the recognition sequence GATC. Therefore, only plasmid DNA would be affected leaving PCR products intact. Since this template contains the kanamycin resistance gene on the cassette, any cells containing this plasmid would render the screening process unreliable. The plasmid pFRT has a size of 3446 bp which implies a strong possibility that one of the forms of this plasmid could coincide with the position of the desired amplification product at 1.7 kb and be undetectable on the gel. The plasmid contains a conditional origin of replication RK6, which requires the Π protein in trans in order to replicate. Although neither of the target strains KRX and BW25113 have *pir+* or *pir116* marker on the genotype, it would still be safe to take into account the rare chance of this plasmid transforming and surviving in the cells. The recombination (see chapter 4.5.11) was carried out according to instructions from the manufacturer (Gene Bridges GmbH, Germany) and the cultures were plated on LB-agar plates supplemented with kanamycin to select for recombinants.

Screening

In the case of *E. coli* BW25113, about 60 colonies were obtained altogether on the LB-Kan plates. The number of colonies were far lower with *E. coli* KRX, but more importantly their reliability was very low since the 'negative control' (no arabinose induction of expression of recombination enzymes) also showed up 2 colonies. All the colonies were further streaked onto LB-Kanamycin plates for screening by colony PCR. A small amount of culture from each streak was dissolved in 30 μL of sterile water and treated in a water bath for 5 min at 98 °C. Later, 2 μl from these cell lysates was used as template in the PCR reaction. In the first set, 10 colonies each were tested from both the strains. A PCR-based screening was designed to conclusively test the presence and positioning of the selection cassette on the genome. The reverse primer Nr. 23 (Table 4.3) bound 200 bases downstream from the reverse flanking homology region while the forward primer O (short for outside; Table 4.3) bound 200 bases upstream of the forward flanking homology region in the genome. The selection cassette was approximately 500 bp longer than the intact *tpiA* gene. An assay was created wherein 3 types of product sizes as shown in Fig. 5.49 were possible:

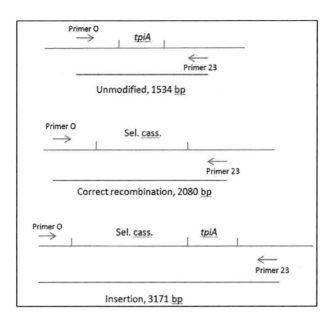

Fig. 5.49: Strategy for the PCR testing of the presence and position of the selection cassette by amplification of a larger portion of the genome.

Figure 5.50 shows the result from this PCR test. As a positive control for the amplification from the genome the 1 kb *leuB* gene was amplified.

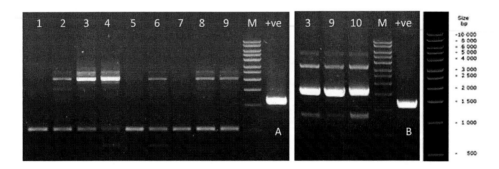

Fig. 5.50: PCR amplification of colonies from *E. coli* BW25113 (A) and *E. coli* KRX (B) based on the strategy to amplify larger genomic regions. '+ve' refers to *leuB* gene amplified as a positive control. M refers to 1 kb molecular size marker from Plasmid Factory GmbH, Germany, which has been shown separately.

In BW25113, the colonies 2, 3, 4, 6, 8 and 9 seemed to show the desired 2 kb band and were thus potentially successful recombinants. In case of KRX, bands were seen at above 3 kb and at 1.5 kb meaning clearly all 3 colonies were either insertions or unmodified cells which somehow had resistance against kanamycin. For all colonies, growth or the absence of it was tested in SGA minimal medium and only colony 6 from BW25113 seemed to be the best choice for further investigation as all others were false-positives.

Complementation of *E. coli* BW25113 (Δ*tpiA*) colony 6 (BW6)

Chemical competent cells of the strain were produced and transformed with plasmid pJET-tpiA and selected on LB-amp plates. From the resultant colonies, three were taken for further analysis. Plasmid isolation was carried out and the DNA observed by electrophoresis to ascertain the presence of the complementing plasmid. These colonies were grown along with the control culture which was the untransformed strain of BW6, in LB with appropriate antibiotics to develop the inoculum. 1-2 mL from the overnight cultures were then centrifuged and resuspended in sterile MilliQ water to be used as inoculum for testing in SGA so that rich media components could not contribute to the growth. The results from the growth comparison are shown in Fig. 5.51.

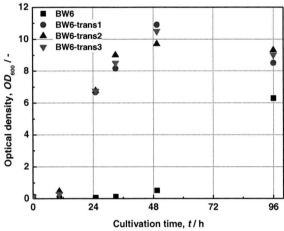

Fig. 5.51: Growth comparison between *E. coli* BW25113 potential Δ*tpiA* (BW6) and 3 transformants with pJET-tpiA in SGA. Shake-flask cultivations were carried out according to the desciptions in chapter 4.6.1.

The transformant cultures clearly showed an advantage in growth in comparison to the untransformed strain. The beginning of growth was after about 10 h and the maximum OD_{600} of about 11 was reached after 50 h whereas the untransformed strain was just beginning to increase in optical density. The final sample taken at an extremely long time of about 90 h showed that biomass growth was significant in the untransformed culture. However, this could have either been due to contaminations introduced during sampling or adaptations by the strain. In any case, a clear growth advantage seen much earlier in the transformant cultures could be taken as a valid sign of complementation. Genomic DNA from the colony BW6 was isolated and primers O and 23 were used to amplify the 2 kb genomic region from Fig. 5.50 resulting in a positive amplification (Fig. 5.52).

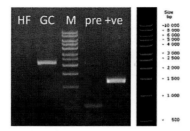

Fig. 5.52: Amplification of 2 kb band for verification of recombination in *E. coli* BW25113Δ*tpiA* (BW6) using the Phusion buffers HF or GC. Pre refers to a previous weak amplification for comparison. +ve refers to an amplification of the gene *leuB* as a positive control. M is 1 kb molecular size marker from Plasmid Factory GmbH, also shown separately for comparison.

Sequencing and confirmation of deletion

The amplified band from Fig. 5.52 was purified and sequenced using the same primers (O and 23). In Fig. 8.8 in the Appendix, the results from the sequencing have been illustrated which clearly showed a perfect to-the-base recombination of the gene *tpiA* with the kanamycin resistance cassette flanked by FRT sites. Thus the strain termed BW6 was *E. coli* K12 BW25113 (Δ*tpiA*) the *tpiA* deletion region of which was exactly the same as the region carried by the complementation plasmid pJET-tpiA (and thus different from the Keio deletion strain).

5.3.8.2 Gene deletion in Strain JM109

The original reference strain *E. coli* JM109 was chosen as the next candidate for gene deletion since such a knockout would act as a good comparison to the reference wild type strain. The reasoning behind making a *tpiA* deletion in the strain JM109 was that, it was the control strain used in the initial studies for β-glucanase expression that showed high levels of enzyme activity when expressed with the plasmid p582 with antibiotic selection pressure. The Keio strain with the *tpiA* deletion and transformed with plasmid pFC4 (p582-*tpiA*) showed complementation of the auxotrophy, but low levels of β-glucanase expression (as described in 5.3.4). In order to check whether the low activity was a feature of the Keio knockout strain (since it also contains a few other deletions), the original control strain was used for the gene deletion.

Transformation of pRedET: Significance of ampicillin concentration

In the recombination experiments mentioned below, the pRedET plasmid with ampicillin resistance was used in order to avoid the inconvenience with light sensitivity of tetracycline. The first step in gene deletion was the transformation of pRedET and maintenance of the plasmid at 30 °C. When ampicillin was applied at the typical working concentration of 100 µg mL^{-1}, a modification of the plasmid by reduction in size was observed (Fig. 5.53).

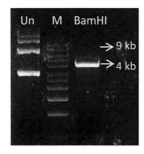

Fig. 5.53: Restriction analysis of plasmid isolated from transformed *E. coli* JM109. 'Un' refers to an unrestricted control. A single cut by BamHI is shown on the lane right to the marker M which is the 1 kb molecular size ladder from Plasmid Factory GmbH with the expected and actual band positions marked.

The unrestricted sample (lane 1) already showed multiple bands that could not be explained as different plasmid forms. A single cut with BamHI (lane 3) showed a linear band at about 4 kb instead of 9 kb. Recombination using such a modified plasmid failed to give any successful knockout colonies, suggesting that the plasmid modification seriously affected the efficiency of expression of the recombination enzymes. In fact, this size reduction was consistently observed in many colonies from a different trial during which again the ampicillin concentration was 'too high' in the medium. A possible explanation was that the low copy number of the plasmid could not tolerate a high selection pressure of ampicillin and created a stress forcing the modification. When the concentration of ampicillin was reduced to 50 µg mL^{-1}, the plasmid structure could be perfectly maintained. Four such transformed colonies were analysed and the corresponding gel photograph is shown below (Fig. 5.54).

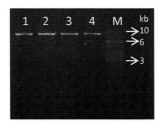

Fig. 5.54: Restriction analysis with BamHI for four colonies after transformation of *E. coli* JM109 with pRedET and cultivation at reduced concentration of ampicillin. Linear bands were seen corresponding to the 9 kb position. 'M' is GeneRuler 1 kb DNA ladder (Fermentas, Lithuania) with the important band positions marked.

Generation of linear cassette and subsequent DpnI treatment

The linear DNA fragment containing the selection cassette for recombination was generated from the template plasmid pFRT by PCR with 2 groups of cycling steps according to the conditions gathered in Table 5.8. The annealing temperature followed a gradient from 56 °C to 61 °C.

Table 5.8: PCR cycling conditions for generation of a selection cassette for homologous recombination.

Step	Temp (°C)	Time (s)
Initial denaturation	98	30
Denaturation	98	10
Annealing 25x	56 → 61	30
Extension	72	80
Denaturation	98	10
Combined annealing + extension 10x	72	110
Final extension	72	600

After the first group of 25 cycles was completed, enough amplicons should have been generated that the primers could bind over their full length and, therefore, the next 10 cycles were operated at a higher annealing temperature combined with extension. The PCR solutions were pooled together and purified, DpnI-treated, and the entire sample loaded onto an agarose gel. Following electrophoresis, the target band was eluted from the gel. The DNA was purified further by sodium acetate precipitation which improved the $A_{260/230}$ absorbance ratio, but resulted in a huge loss, so that the final concentration of the linear cassette was 24 ng μL^{-1}. This preparation was used for recombination which lead to a number of colonies (see chapter 4.5.11). Two of such potential knockout colonies were tested for the genotype with the correct recombination event. The PCR test with primers O and 23 was performed as described above and the result is seen in the following gel photograph (Fig. 5.55).

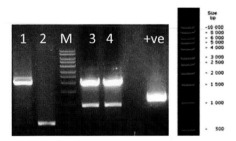

Fig. 5.55: PCR test to check the genomic region of two potential recombinant colonies of *E. coli* JM109. Lanes 1 and 2 show PCR reactions from lysates of colonies 1 and 2, respectively. Lanes 3 and 4 show PCR reactions from isolated genomic DNA of colonies 1 and 2, respectively. The lane +ve is an amplification of gene *leuB* as positive control M refers to 1 kb DNA ladder from Plasmid Factory GmbH, which is also shown separately.

The results showed that the required 2 kb amplicon was seen in the PCR reactions from genomic DNA of the 2 colonies. In a subsequent preparative PCR, the 2 kb band was amplified from the genomic DNA in a large amount, purified from gel and sequenced using the same primers O and 23. The result of the sequencing shown in Fig. 8.9 in the Appendix was a definitive proof that the recombination was successful in both the colonies. However, upon closer analysis of the sequence in Fig. 8.9, image B, it was found that about 15 bases were missing from the promoter of the kanamycin resistance gene inserted into the genome. But the fact that these colonies were repeatedly tested to be kanamycin-resistant, proved that the loss of these bases had not in any way affected the expression of the resistance gene. The two colonies were stored as glycerol stocks, designated JM1 and JM2, and represented *E. coli* JM109 ($\Delta tpiA$). To test the *tpiA* auxotrophy on the phenotypic level, three colonies each from the strains JM1 and JM2 transformed with the complementation plasmid pFC4 were cultivated in shake-flasks in SGA medium and compared to the knockout strains without plasmid. All the transformed strains showed growth, albeit with different growth characteristics as seen in Fig. 5.56.

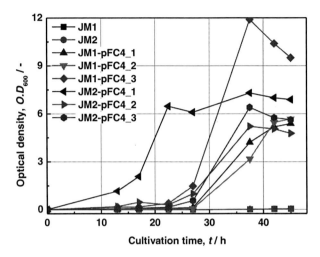

Fig. 5.56: Growth profiles from shake-flask cultivation of knockout strains JM1 and JM2 and their transformant colonies in SGA medium. The cultivations were antibiotic-free and carried out at 37 °C. Other general conditions were according to the descriptions in chapter 4.6.1.

The growth test proved the ability of the plasmid pFC4 to complement the auxotrophy in the knockout strains JM1 and JM2. These two strains showed no growth without the plasmid throughout the time period of the experiment. Furthermore, plasmid isolation and restriction digestion with appropriate enzymes was performed on all the transformed strains and the presence and identities of the plasmids were confirmed.

Information about the pFRT template plasmid

One extremely limiting step in the recombination experiments was the quantity and purity of the linear DNA fragment containing the kanamycin resistance cassette. The template for generating this cassette by PCR was provided by the manufacturer (Gene Bridges GmbH, Germany) in the form of a plasmid that cannot be transformed into routine strains since it contained an R6K conditional origin of replication. These plasmids require the presence of the *pir* marker on the genome which codes for the Π protein in order to replicate. Two such strains are *E. coli* BW25141 (*pir+* marker) and *E. coli* BW23474 (*pir116* marker for high copy number). The template plasmid pFRT was transformed into the strain BW23474 and verified by restriction digestion of the isolated plasmid (Fig. 5.57). These strains would ensure a reliable supply of plasmid material in order to optimize the PCR and generate an adequate quantity of linear amplicon for recombination.

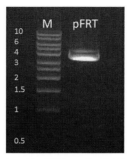

Fig. 5.57: Identification check of the plasmid pFRT restricted (single cut) with the enzyme NotI (Image courtesy: Wünsch, Fermentation Engineering, Bielefeld University). Molecular sizes of the bands for the standard ladder M (Plasmid Factory GmbH) are given in kb.

Characterization of growth and β-glucanase expression with *E. coli* JM2-pFC4

One of the strains of JM2-pFC4 shown in Fig. 5.56 was tested in batch mode in the in-house fermenter with 1 L working volume in SGA medium at a temperature of 37 °C without antibiotics to check if the activity levels reached with the control strain could be achieved again using this auxotrophic system (Fig. 5.58).

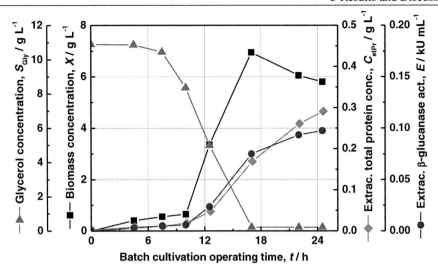

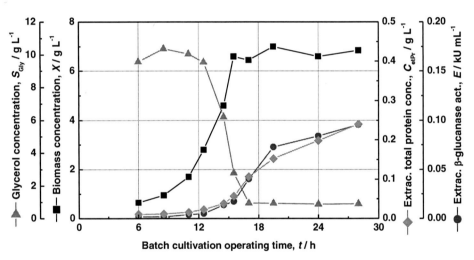

Fig. 5.58: History of two batch cultivations of JM2-pFC4. The experiments showed almost identical results. General cultivation conditions were as described in chapter 4.6.2 for cultivations with the 2 L in-house fermenter. Since the strain JM2-pFC4 was derived from the reference strain *E. coli* JM109, the process temperature was set at 37 °C.

From the batch experiment (Fig. 5.58) it was clear that this auxotrophic system could grow normally in SGA medium to biomass concentrations of 7 g L^{-1}, but the maximum recombinant β-glucanase activity was nowhere near the levels reached with the control strain (96 U mL^{-1} against 1320 U mL^{-1}). In fact, the activity reached with this system was on the same scale as that with the Keio strain expressing this plasmid (see chapter 5.3.4). The maximum total protein concentration in the extracellular medium was 0.23 g L^{-1} and the ratio of the activity to the protein concentration was found to lie at 0.39 kU mg^{-1} which was even lower than the value of 0.5 kU mg^{-1} discussed previously for JW3890-2- pFC4 (chapter 5.3.4).

5.4 Efficiency of various strains and plasmids

To conclusively check the construct pFC2, it was transformed into standard strains such as *E. coli* JM109 and DH5α. At the same time, the knockout strain *E. coli* JW5807-2 also had to be tested and it was transformed with the original plasmid p582, which is known to give high activity levels in the reference strain *E. coli* JM109. These cultures were then grown in shake-flasks in SGA at 37 °C and the data for biomass and extracellular β-glucanase activity is shown in Fig. 5.59.

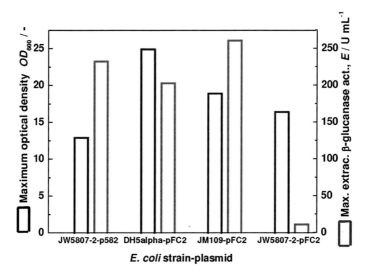

Fig. 5.59: Maximum optical density and activity of β-glucanase in the extracellular fraction from shake-flask cultures of different strains harbouring plasmid p582 or pFC2 cultivated at a temperature of 37 °C.

In the strain JW5807-2, plasmid p582 was maintained with ampicillin since the strain already had kanamycin resistance in the genome. For *E. coli* DH5α and *E. coli* JM109 transformed with the construct pFC2, the maintenance was achieved with kanamycin while in the case of strain JW5807-2-pFC2, no antibiotics were used and the maintenance was based on dependence on *leuB*. From Fig. 5.59 it can be seen that the plasmid pFC2 was actually capable to express the recombinant β-glucanase at a reasonable level (250 U mL^{-1}) when using the usual *E. coli* strains like JM109. The Keio knockout strain JW5807-2 was also performing well with the original plasmid p582 and gave an activity of comparable maximum level. However, the combination of knockout strain and complementation construct pFC2 proved to be inefficient. The enzyme activity found in this case was extremely low (10.5 U mL^{-1}) even though the biomass growth was comparable to the other strains. It did not seem possible that the plasmid was lost due to instability since, the growth data showed otherwise, and the plasmid was essential for growth in SGA. Therefore this problem could be explained as an effect of the temperature of 37 °C and the associated formation of inclusion bodies which was discussed in chapter 5.2.3.

The control strain *E. coli* JM109 transformed with the complementation plasmid pFC2, pFC4 or the reference plasmid p582 were grown in shake-flasks in SGA medium at 30 °C. For comparison, the reference strain JM109-p582 was also grown at 37 °C (Fig. 5.60). All flasks contained kanamycin for selection pressure. Due to non-optimal biomass growth and a possible discrepancy in the enzyme assay, the activities for all the strains were measured to be lower than previously measured levels. Nevertheless, these values allowed comparison to the control and estimation of the characteristics of these strains.

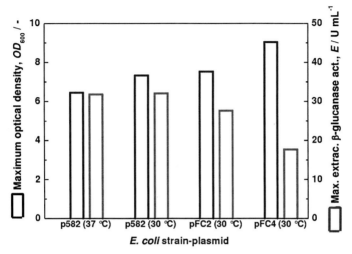

Fig.5.60 Influence of process temperature on the cultivation of *E. coli* JM109 transformed with various plasmids.

The results in Fig. 5.60 show that the strain JM109 with p582 was the ideal case with maximum β-glucanase activity. The construct pFC2 in the same host delivered only a slightly lower activity while the construct pFC4 with the *tpiA* gene delivered an activity of about 50%. Thus, the constructs inherently had a characteristic of reduced target enzyme activity due to the increased size and/or the expression of the complementation gene.

5.5 Comparison of reference and alternative systems

5.5.1 Comparison of β-glucanase expression

Finally, a rough representative diagram (Figure 5.61) is presented that compares the efficiencies of different systems: strain-plasmid combinations, processes, reactors *etc.* The different levels of extracellular recombinant β-glucanase activities achieved using each system is shown. The different scaling of the graphs, set in order to improve visibility, needs to be considered. From this summary of results from various experiments, it is seen that the *leuB* complementation system JW5807-2-pFC2 came closest to the benchmark set by the reference strain in terms of expression of the recombinant β-glucanase. Although antibiotic-free cultivation was possible in the *tpiA* complementation system, extracellular activities of β-glucanase were much lower.

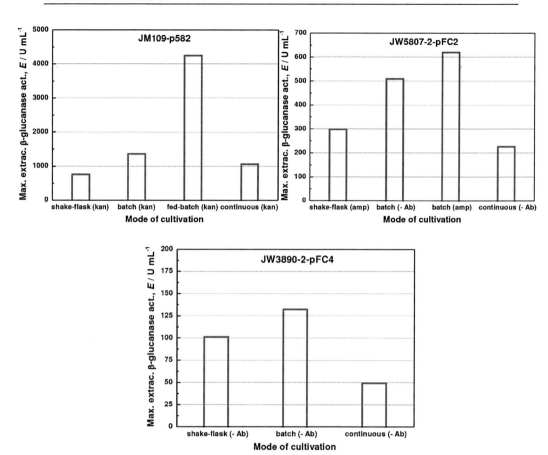

Fig. 5.61: Comparison of maximum extracellular β-glucanase activities by the reference system *E. coli* JM109-p582 and the two auxotrophic systems *E. coli* JW5807-2-pFC2 and *E. coli* JW3890-2-pFC4. '-Ab' refers to absence of antibiotics while 'amp' and 'kan' refer to presence of ampicillin and kanamycin respectively.

5.5.2 Comparison of plasmid selection efficiency

To substantiate the results, long-term plasmid segregational stability was studied for various strains by means of a plating and colony picking method (chapter 4.7.10). Continuous cultivation was performed in SGA medium with glycerol at 20 g L^{-1} concentration with or without the presence of antibiotics. For the strain *E. coli* JW5807-2-pFC2, the stability was close to 100% over a period of 200 h of continuous cultivation without the addition of any antibiotics. The strain *E. coli* JW3890-2-pFC4 showed quite a lot of fluctuation in the fraction of plasmid-carrying cells, but was still able to recover back to near maximum levels by the end of 205 h. In comparison, the reference strain *E. coli* JM109-p582, was reduced to a population of 79% of plasmid-carrying cells after 190 h, in the absence of selection pressure. Intriguingly, the plasmid p582 in the absence of antibiotic selection pressure was not as highly unstable in the reference strain as previously thought. While the culture with antibiotic maintained near 100% retention of plasmid in the cells, the antibiotic-free culture lost the plasmid only to a small extent (Fig. 5.62).

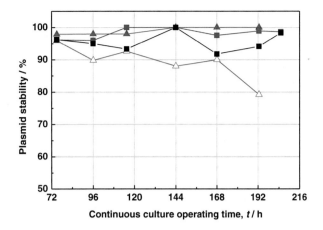

Fig. 5.62: Plasmid stability comparison between the reference strain *E. coli* JM109-p582 in the presence (closed triangles) and absence (open triangles) of antibiotic selection pressure and the complemented auxotrophic strains *E. coli* JW5807-2-pFC2 (blue closed squares) and *E. coli* JW3890-2-pFC4 (black closed squares) in the absence of antibiotic selection pressure. Experimental procedures have been described in chapter 4.7.10.

Segregational plasmid stability remains an area of serious concern for industrial recombinant processes which makes it essential to develop strategies for improving plasmid stability. For example, the *infA* complementation system reported by Hägg *et al.*, (2004) showed good segregational plasmid stability in antibiotic-free culture over many generations in comparison to the parent strain as control.

Particularly, plasmids expressing high levels of recombinant proteins are prone to be selected against. The effect of high transcription and translation rates on the specific plasmid loss rate of pBR322-derived plasmids was shown recently (Popov *et al.*, 2011). Although plasmid p582 is based on the pUC19 origin of replication and is therefore expected to have a high copy number, the segregational stability was found to decrease over long periods of cultivation in the absence of selection pressure from kanamycin. This characteristic of plasmid p582 could explain how 79% of the population still carries the plasmid even without selection pressure for 192 h. The probability of segregation for a high-copy number plasmid (discussed in chapter 2.4) should have worked in favour of the stability resulting in the situation where it was quite 'difficult' to generate a plasmid-free cell by chance. It also seems that the plasmid-free cells that are actually generated in the antibiotic-free culture, can not really show up a significant growth rate advantage against the plasmid-carrying cells. The fact that p582 is of only moderate size (6.1 kb) and the recombinant protein was not highly toxic to the cell are other aspects that may have aided in the stability. Another aspect is the difference seen between the decline in isolated plasmid concentration and decline in plasmid segregational stability for the reference strain in the absence of antibiotics (Figs. 5.8 and 5.62). The decline in isolated plasmid concentration reflected the absolute loss of plasmids through segregation and thus also included loss in copy number in a fraction of the population. On the other hand, the plating and colony-counting method gave an idea of the whole population, since even cells with reduced copy number could still turn up positive as a resistant colony when streaked on antibiotic plate. Therefore, the decline was not very drastic as seen with the decline in isolated plasmid content. Nevertheless, the plasmid stability of the auxotrophic systems in Fig. 5.62 showed near 100% levels throughout without depending on antibiotics and therefore, represented an improvement over the reference system.

5.6 Analysis of P_{fic} activity by RT-qPCR

In the continuous culture experiments described previously in chapters 5.1 to 5.3, one of the most important aspects was the regulated expression of the *kil* gene coding for BRP in order to realize extracellular target enzyme expression. Therefore, during operation of the chemostat at different specific growth rates of the strain, the regulation of the stationary-phase promoter P_{fic} controlling *kil* expression takes an important role. A closer analysis of the activity of this promoter is therefore of interest to derive information as to how the target protein export to the medium is regulated during a chemostat and thus make it possible to efficiently guide optimizations in the process for the goal of improving the enzyme productivity. Whereas other techniques such as microarray analysis and next-generation tools like RNA-sequencing are suitable for transcriptome-wide investigations, RT-qPCR with its accuracy, efficiency and sensitivity remains the gold standard for analysing a small number of transcripts from a large sample pool or even for medium throughput gene expression analysis (Derveaux *et al.*, 2010; Perkel, 2013).

Since P_{fic} is a part of the regulon of the stationary-phase sigma factor RpoS, P_{fic} activity could be compared with the expression of the *rpoS* transcript. Additionally, comparison with the transcript levels of the primary sigma factor *rpoD* should be able to give a picture of the activation of P_{fic} with respect to the metabolic status of the cell. Therefore, the activity of the promoter P_{fic} over different growth rates was to be analysed by means of quantification of the *kil* mRNA transcripts through reverse transcription-quantitative real-time PCR and comparing with the change in expression pattern of the stationary phase sigma factor gene *rpoS* and the primary sigma factor gene *rpoD*. The aim was to find if there would be a significant difference in the absolute mRNA level between different steady states for the target genes – *kil*, *rpoS* and *rpoD*.

5.6.1 Primer functionality

Forward and reverse primers designated as left and right for the respective target gene (Table 4.3) were designed to amplify short fragments within the coding sequences of these genes for use in real-time PCR. Apart from the three main targets *kil*, *rpoS* and *rpoD* which would be quantified, primers were also designed for other targets such as *bgl*, *leuB*, *tpiA*, *tetR* and *gapA* for initial comparisons. The functionalities of the primers were checked in an end-point PCR using Phusion polymerase. PCR amplification of 30 cycles was performed according to the descriptions in chapter 4.5.3. A common annealing temperature of 55 °C was used for all targets and the extension step was reduced to 12 s. For the *kil* gene, the plasmid p582 or pFC2 was used as template, whereas genomic DNA from *E. coli*

JW5807-2 was used as template for the genes *rpoS* and *rpoD*. The amplicons were analysed in an agarose gel to confirm if the amplification was efficient and specific (Fig. 5.63).

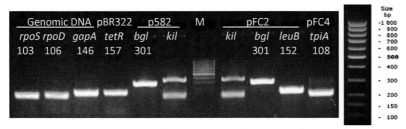

Fig. 5.63: Result of initial PCR experiment for testing primer functionality. The templates used are given above and the target genes and amplicon sizes in base pairs shown below. M refers to 100 bp DNA ladder from Plasmid Factory GmbH which is also shown separately. The brightest band in the ladder corresponds to 500 bp.

The additional target *gapA* was amplified from genomic DNA of *E. coli* JW5807-2, the tetracycline resistance *tetR* from plasmid pBR322, the recombinant β-glucanase gene *bgl* from p582 or pFC2, complementation gene *leuB* from pFC2 and the complementation gene *tpiA* from pFC4. The target sequence in the *kil* gene had been initially amplified with a different set of primers which resulted in non-specific amplification and prompted a fresh design of primers and renewed testing. From Figure 5.63, it could be seen that with the exception of *kil*, all other bands were specific and of the expected amplicon size according to the primer design. New primers, given in Table 4.3 were designed for the *kil* target using the *kil* gene present on plasmid pFC2. This resulted in a single amplicon of the expected size (shown later in Fig 5.65).

Another objective from primer design for all targets was to prove the primer specificity for the plasmid-borne and the genomic genes. The possibility of non-specific binding of the primers to *E. coli* BW2952 genomic DNA sequence was ruled out by *in silico* analysis (chapter 4.5.10). The primers for *kil* gave no matches while *rpoS* and *rpoD* gave unique matches (1 genomic copy). Furthermore, primers for genomic genes *rpoS, rpoD* and *gapA* were tested in an end-point PCR with 2 different templates – genomic DNA from *E. coli* JW5807-2 and from plasmid pFC2. In the agarose gel image shown below (Fig. 5.64), a clear amplification was seen in the case of the genomic DNA template (lanes 1-3), while faint amplification was also seen with the pFC2 template (lanes 4-6). This observation could be due to some genomic DNA impurities in the plasmid preparation and the low yield of the observed product supports this possibility. Lane 7 (No template control) showed typical elongated primers in the absence of any template.

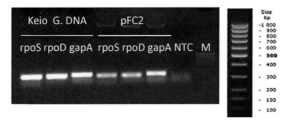

Fig. 5.64: End-point PCR to check primer specificity for genomic targets *rpoS*, *rpoD* and *gapA*. M refers to 100 bp DNA ladder from Plasmid Factory GmbH shown separately.

In the next experiment, the primers for the plasmid-borne genes *kil*, *bgl* and *leuB* were tested using the same templates as previously. In the result shown in Fig. 5.65, the primers for *bgl* and *leuB* showed no amplification with genomic DNA from *E. coli* JW5807-2. However, a slight amplification was seen in the case of *kil*. Hence, the *kil* primers seemed to bind weakly at some position in the genome creating an amplicon of comparable size to the target amplicon that was produced from the plasmid pFC2 as template (lane 4). This necessitated the treatment with DNase during RNA purification to avoid ambiguous results from genomic DNA contaminations. The new primers designed for the *kil* target (Table 4.3) gave a specific amplicon of expected size of 97 bp.

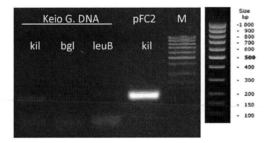

Fig. 5.65: End-point PCR to check primer specificity of plasmid-borne genes *kil*, *bgl* and *leuB*. M refers to 100 bp DNA ladder from Plasmid Factory GmbH which is also shown separately with sizes in bp.

5.6.2 Preparation of plasmid DNA standards for absolute quantification

Through absolute quantification, it would be possible to find out the absolute number of copies of *kil*, *rpoS* and *rpoD* mRNAs present in any sample through direct readout from their respective standard C_T curves. Therefore, it was necessary to have preparations with known copies of these fragment molecules. The most precise standards for absolute quantification in RT-qPCR would be known standards of the same RNA molecule that is analysed, since this would then take into account the effect of variable efficiency of reverse transcription. This would require specific mRNA molecules to be generated through *in vitro* transcription and quantification by spectrophotometry. To avoid such a procedure that offered more possibilities of error due to RNase contamination, the reverse transcription process is assumed to cause no significant limitations in this study. Therefore DNA standards in the form of the cloned constructs were used for standardization.

For generating standard curves for genomic genes like *rpoS* and *rpoD*, their target regions had to be cloned onto a standard vector. The corresponding primers from Table 4.3 were used in a conventional PCR with Phusion polymerase to create blunt-ended amplicons that were cloned onto the pJET1.2 vector backbone (Thermo Scientific, Germany) thus giving rise to the plasmids pJET-rpoS-frag. and pJET-rpoD-frag (Fig. 5.66). The transformants were screened by PCR using the same primers as for the cloning (respective left and right primers from Table 4.3). As negative control, the pJET blunt vector was used as template for the amplification. In the case of *rpoS*, all 4 clones tested were positive and they showed the band corresponding to 100 bp which was the size of the target amplicon. The negative control lane showed no amplification. In the case of *rpoD*, clones 2 to 4 possessed the fragment of the correct size (100 bp). The negative control seemed to suggest an amplification but this band was absent in the lane for clone 1 and is therefore probably only a result of a non-specific interaction between the primers and the pJET backbone.

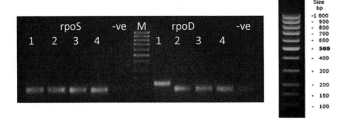

Fig. 5.66: Screening for clones of pJET-rpoS-frag. and pJET-rpoD-frag. –ve refers to the negative control. M refers to 100 bp molecular mass ladder from Plasmid Factory GmbH which is also shown separately for comparison (not to scale).

Although the *kil* gene was already present on a plasmid, to maintain uniformity, this target region was cloned similarly onto pJET resulting in the plasmid pJET-kil-frag. These plasmids were maintained in strains *E. coli* JM109 or *E. coli* Top10. From a routine plasmid isolation followed by measurement of the concentration of DNA, the copies of the target fragment could be calculated by the relation described in chapter 4.8.4. This allowed the determination of the number of copies of any target molecule added to a qPCR reaction.

5.6.3 RNA isolation and quality control

The reference strain *E. coli* JM109-p582 was cultivated in a chemostat at 37 °C with SGA medium containing thiamine and with selection pressure from kanamycin (chapter 4.6.3). Nine samples were taken during the batch-phase of growth, whereas in the chemostat phase, samples were collected from space velocities of 0.05 h^{-1}, 0.1 h^{-1}, 0.2 h^{-1}, 0.3 h^{-1} and 0.4 h^{-1}. For each space velocity, samples were taken at 2 different times separated by an average of 10 h and each sample was taken in duplicate. Sample processing, RNA isolation and off-column DNase treatment were done as described in chapter 4.8.2. Using the RNA Protect Bacteria Reagent (QIAGEN GmbH, Germany), the samples were processed so that the intracellular RNA remained intact and RNA isolation was performed on all of the samples simultaneously.

5.6.3.1 Concentrations and purity

The following table (Table 5.9) gives the concentrations and purity ratios of the RNA preparations from the samples taken during the initial batch growth phase (before feeding). The values given here are after an off-column DNase-treatment on the RNA preparations.

Table 5.9: Concentration and purity of RNA of samples from the batch-phase of a continuous cultivation with *E. coli* JM109-p582, measured with the Nanodrop spectrophotometer.

S.No.	Sample time (h)	Concentration (ng μL^{-1})	Ratio A_{260}/A_{280}	Ratio A_{260}/A_{230}
1	7	995	2.1	2.2
2	9	783	2.09	0.32
3	11	548.1	2.09	0.51
4	13	45.5	2.36	0.78
5	15	257	2.18	2.07
6	17	224	2.14	0.69
7	19	83.2	2.22	1.28
8	23	4.6	8.41	0.03
9	24	25.7	2.16	1.13

With the exception of the sample from 23 h, all other RNA preparations had a reasonable concentration that would allow uniform transcript analysis. The same pattern was also seen for the A_{260}/A_{280} purity ratio, where the extremely low values for both nucleic acid and protein probably resulted in the ratio of 8.41, whereas all other preparations had very good purity values (about 2.0). The A_{260}/A_{230} ratios for some of the preparations were very low (less than 1.0), and this was probably a result of the clean-up procedure after the off-column DNase treatment, not having been able to completely eliminate all chaotropic salts.

The following table (Table 5.10) gives the concentrations and purity ratios of the RNA preparations of samples from the chemostat-phase after off-column DNase treatment.

Table 5.10: Concentration and purity of RNA samples from chemostat-phase measured with the Nanodrop spectrophotometer. Sample ID B1T1 refers to biological replicate 1, technical replicate 1.

S. No.	Space velocity (h^{-1})	Sample ID	RNA concentration (ng μL^{-1})	Purity ratio A_{260}/A_{280}	Purity ratio A_{260}/A_{230}
1	0.05	B1T1	86	4.8	1.5
2		B1T2	30.4	4.22	2.11
3		B2T1	27.6	2.92	0.54
4		B2T2	29.6	3.4	0.91
5	0.1	B1T1	166	3.0	1.3
6		B1T2	145.5	3.1	0.26
7		B2T1	27.4	2.94	0.22
8		B2T2	59.2	2.13	1.92
9	0.2	B1T1	72	2.1	0.69
10		B1T2	59.3	1.92	1.37
11		B2T1	60.8	2.12	0.12
12		B2T2	113.6	2.05	0.26
13	0.3	B1T1	102	1.98	0.31
14		B1T2	204.5	2.12	2.3
15		B2T1	110.7	2.09	2.21
16		B2T2	161.2	2.13	0.44
17	0.4	B1T1	193.6	2.09	2.1
18		B1T2	118.5	2.07	1.31
19		B2T1	235.5	2.12	1.81
20		B2T2	295.8	2.11	1.13

The RNA concentration and purity values for the chemostat-phase samples were quite similar to those seen for the batch-phase samples. The A_{260}/A_{280} purity ratio was always about 2.0 though for some samples it was as high as 4.0, probably as a result of extremely low absorbance at 280 nm.

5.6.3.2 Electrophoresis

Though normally for RNA electrophoresis, a denaturing agarose gel containing formaldehyde would be required in order to resolve the secondary structures and to properly correlate size and distance travelled, a simple non-denaturing 2% agarose gel electrophoresis was carried out to check the quality of the RNA. Generally, intact total RNA electrophoresed on an agarose gel should show a 2:1 ratio of the 23S to 16S rRNA band intensities. The electrophoresis analysis of the RNAs from chemostat-phase samples after the off-column DNase treatment is shown in Fig. 5.67. Almost all the RNA preparations showed the expected bands corresponding to the 23S and 16S rRNA. Although, a few lanes showed a slight streaking of the bands, no major degradation of RNA was to be found. Therefore, the measures taken against RNase contamination could be stated to have been effective.

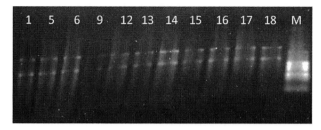

Fig. 5.67: Electrophoresis of selected RNA preparations from chemostat-phase samples numbered according to Table 5.10. Standard Low Range RNA ladder SM1831 (Fisher Scientific-Germany GmbH) loaded on lane M served only as a general reference to compare for intact RNA bands on an agarose gel.

5.6.3.3 Absence of genomic DNA contamination

The significance of this characteristic was found after initial trials in which contaminating DNA molecules probably offered extra copies of the target that seriously affected the reliability of the quantification of mRNA transcripts. Although an on-column DNase treatment step was integrated into the RNA isolation procedure, it became clear that it was not sufficient and an extra off-column DNase treatment was required on the freshly isolated RNA preparations to be able to convincingly remove most of the genomic and plasmid DNA contaminations. Such an off-column DNase treatment step on RNA preparations was also reported by Leong *et al.*, (2007) who found that additional higher molecular mass amplicons from genomic DNA clearly showed up only in samples not treated by DNase I digestion. Primers for RT-qPCR from eukaryotic samples could be designed such that they bind over an intron-exon boundary thus minimising the chances of binding to genomic DNA. Obviously, this advantage is not available for *E. coli*. The test for genomic DNA contamination was carried out as described under chapter 4.8.3. Firstly, the comparison of the threshold fluorescence cycles for RNA isolation using only on-column DNase treatment is shown in Fig. 5.68.

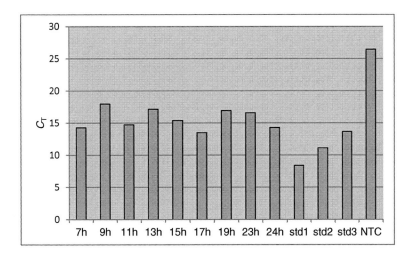

Fig. 5.68: Threshold fluorescence cycle values of RNA from batch-phase samples for quality control in a No-RT experiment after performing only an on-column DNase treatment step. Samples are according to the batch sample times in Table 5.9. Experimental procedures are detailed in chapter 4.8.3.

The comparison of C_T values showed that the samples were very far-placed from the ideal level of the NTC, which meant that they still contained a substantial amount of DNA contamination. Some samples had very high DNA contamination that their C_Ts matched that of Std3. The standards were serially diluted genomic DNA preparations and showed a clean C_T difference of about 3 cycles (one cycle = doubling; 3.3 cycles = 10 times increase). Performing an off-column DNase treatment on the isolated RNA preparations and repeating the No-RT quality control experiment resulted in a distinct shift of the C_T values towards that for the NTC as seen in Fig. 5.69.

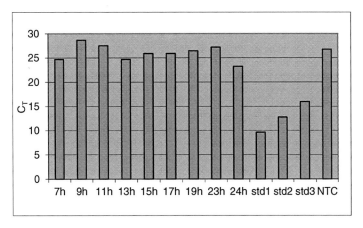

Fig. 5.69: Threshold fluorescence cycles in a No-RT experiment after application of an off-column DNase treatment step on the isolated RNA preparations from batch-phase samples. Samples are according to the batch sample times in Table 5.9. Experimental procedures are detailed in chapter 4.8.3. See Fig. 5.68 for details.

Similarly, for the isolated RNAs from the chemostat-phase samples after the off-column DNase treatment, a similar No-RT experiment was carried out. In this case, 4 serial dilutions of the standard genomic DNA were applied (up to 0.54 ng μL^{-1}). The results seen in Fig. 5.70 show once again the successful elimination of almost all DNA in the RNA preparations, since their C_T correlated with that for the NTC.

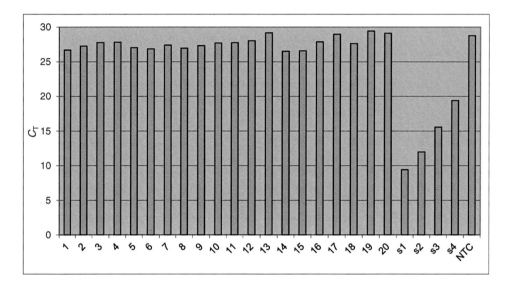

Fig. 5.70: Threshold fluorescence cycles in a No-RT experiment after application of an off-column DNase treatment step on the isolated RNA from chemostat-phase samples. Samples are according to the serial numbers in Table 5.10. Experimental procedures are detailed in chapter 4.8.3.

The analysis of the melting curve (Fig. 5.71) for these samples throws more light on the nature of the contaminating DNA molecules. A unique sharp curve indicated a specific amplicon (target) whereas multiple peaks indicated parallel amplifications. Figure 5.71 shows the melting peaks for the standard genomic DNA (s1-s4) which occur at about 82.5 °C and the peak for the amplicons resulting from primer interactions which occur at a lower temperature of around 74 °C. It is seen that a few of the RNA samples show a slight change in fluorescence corresponding to the higher temperature which points to a small amount of residual DNA in the preparation. This resulted in a characteristic double peak with one high -dF/dT peak at a lower temperature corresponding to the primer-dimers but additionally a smaller -dF/dT peak at a high temperature pointing to a small fraction of amplicons from genomic DNA.

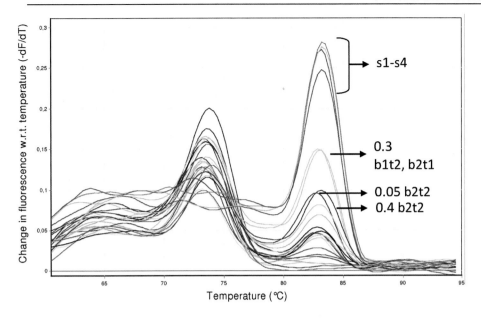

Fig. 5.71: Melting curve analysis of amplicons from a No-RT quality control experiment using RNA from chemostat-phase samples after off-column DNase digestion (refer to Table 5.10).

5.6.4 Batch-phase transcript profiles for *kil*, *rpoS* and *rpoD*

The volume of RNA preparation added to the 1-step RT-qPCR was normalized to the concentration from Table 5.9. Each transcript was analysed in each sample in a double estimation along with its standards from a single master mix preparation. The data presented in the subsequent sections give the quantified target transcripts as copies per RT-qPCR reaction. Related to the amount of total RNA taken as template for each analysis, the data represent the number of copies of the target present in 20 ng of total RNA isolated from the cells of each sample. Although the same number of cells was taken for each sample for the isolation of RNA (refer to 4.6.2), the RNA yield from each sample could not be achieved constant (Table 5.9). Therefore, due to the normalization with respect to the amount of template RNA, the calculated target copies represented a fairly accurate profile of the amounts of transcripts of each target present in the cells for any given sample.

Transcription of *kil*

One-step RT-qPCR was performed using the Rotor-Gene SYBR Green RT-PCR kit (QIAGEN GmbH, Germany) according to the reaction setup shown in Table 5.11. Both reverse transcription of RNA templates and subsequent real-time PCR amplification and detection of target amplicons took place in one reaction vessel and within a single program. In contrast to this, a two-step method would require cDNAs generated previously to be taken from the reverse transcription reaction and added to the real-time PCR amplification tubes, thus greatly increasing chances of RNase contamination during handling. The reverse primer (kil-right) functioned additionally as a specific primer for complementary DNA generation during the reverse transcription step.

Table 5.11: Reaction setup for RT-qPCR of *kil* target amplicons of RNA from batch-phase samples. The volumes given for template RNA and RNase-free water are schematic. These volumes were variable and depended upon the RNA concentrations shown in Table 5.9, such that 20 ng of RNA was taken per reaction.

Component	Volume per reaction (µL)	Final concentration
2x Rotor-Gene SYBR Green RT-PCR Master Mix	12.5	1x
Primer kil-left (50 µM)	0.5	1 µM
Primer kil-right (50 µM)	0.5	1 µM
Rotor-Gene RT Mix	0.25	
Template RNA	1	20 ng per reaction
RNase-free water	10.25	
Total	**25**	

The RT-qPCR analysis was performed according to the program shown in Table 5.12. After the initial incubation for reverse transcription, the samples could be directly analysed further within the same run after a short activation of the Taq DNA polymerase.

Table 5.12: Cycling conditions for one-step reverse transcription and real-time PCR amplification of RNA from batch-phase samples.

Reaction step		Temperature (°C)	Time (s)
Reverse transcription		55	600
Taq polymerase activation		95	300
Cycling (40x)	Denaturation	95	5
	Combined annealing + extension	60	15*

* Fluorescence data acquisition performed at the end of extension in each cycle

In Fig. 5.72, the amplification profiles (in double) for the RNA samples and NTC are shown. The normalized fluorescence curves showed no signs of any irregular changes. The NTC had a high C_T of 30.63 which was very ideal and indicated weak interaction between the primers. The duplicate estimations for the samples coincided very closely with each other. One standard reaction was run with a given concentration of 1.01×10^7 copies per reaction. The C_T from this standard was used to adjust the imported standard curve which was generated in a separate trial with serial dilutions of pJET-kil-frag. The advantage of this feature in the software (Rotor Gene, Corbett Research) was that, for one set of primers, target and master mix solution, the standard curve needed to be determined only once. This curve could then be imported in future analyses of unknowns of the same target with the same primers. Along with the unknowns, only one known standard needed to be analysed, which was used by the software as a marker to calibrate the imported standard curve. The efficiency of the imported standard curve was retained whereas the y-intercept was adjusted according to the number of cycles required until the reactions became visible in the current run.

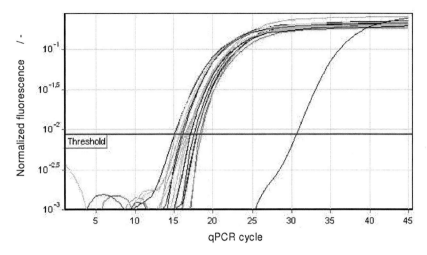

Fig. 5.72: Normalized fluorescence curves from the amplification of reverse-transcribed *kil* targets in a RT-qPCR experiment. The threshold fluorescence level of 0.0088 was set manually within the exponential phase of amplification.

Standard curve

A standard curve was generated from serial dilutions of the pJET-kil-frag. plasmid preparation ranging from concentrations of 33.6 ng μL^{-1} to 33.6×10^{-6} ng μL^{-1} corresponding to 1.013×10^{10} to 1.013×10^{4} copies of *kil* target per reaction (C_r), respectively (Fig. 5.73). For this qPCR experiment based on DNA templates, the Rotor Gene SYBR Green PCR kit was used.

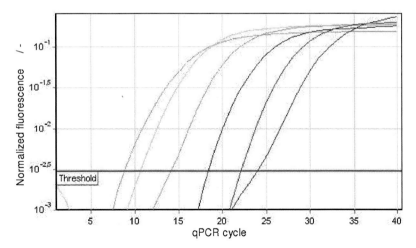

Fig. 5.73: Normalized fluorescence curves from standard plasmid DNA amplification for *kil* targets. The threshold fluorescence level of 0.003 was set manually within the exponential phase of amplification.

The normalized fluorescence curves were found to be separated regularly at the expected intervals of approximately 3 cycles per 10x dilution. Normalized fluorescence refers to the dynamic normalization performed on each sample by the software (Rotor Gene, Corbett Research) which calculated the average background fluorescence for each sample from the fluorescence collected at the starting cycle to the actual cycle just before amplification commenced. The fluorescence data for any sample was then divided by its average background fluorescence and thus the actual fluorescence increase from amplification was normalized to give precise quantification data. This was displayed on a logarithmic scale in order to focus closely on the most relevant amplification portion of the experiment. Although originally, 7 orders of magnitude were included, the sample with the highest concentration had to be neglected from the standard analysis due to a highly irregular amplification. The standard curve thus generated is shown in Fig. 5.74. This curve plots C_T versus logarithm of target copy concentration (C_r) and is represented by the linear equation

$$C_T = -3.273 \times \lg(C_r) + 37.614 \tag{49}$$

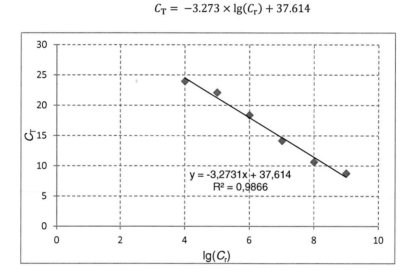

Fig. 5.74: qPCR standard plot for known concentrations of the *kil* standard plasmid pJET-kil-frag.

Logarithm to the base 10 was used in these calculations since it is practical for preparation of serial dilutions of known standards. The standard plot gave an apparent efficiency of amplification (E_x) of 1.02 and a slope (m) of –3.27. The value of the y-intercept is an indicator of the sensitivity of the qPCR assay. Low values of y-intercept (theoretically, the C_T corresponding to the least template concentration) indicate high sensitivity of the assay (Bustin, 2000). A negative slope with a magnitude lower than the ideal of 3.322 as seen here gives an apparent efficiency higher than 1.0 but actually indicates the presence of inhibitors in the reaction (Real-Time PCR Brochure, Qiagen Resource Center). Based on the standard equation, the transcript copies for the *kil* target were calculated in duplicate and the average and standard deviations computed as well. The levels of the *kil* transcript were overlaid upon the batch growth curve to derive a complete profile of the changes in this target over the batch growth phase for equal amounts of total RNA (Fig. 5.75).

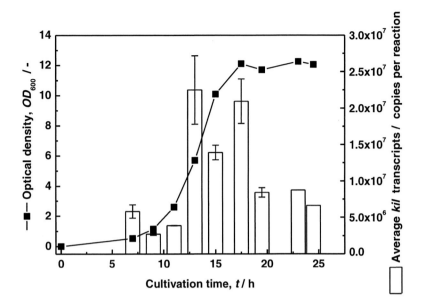

Fig. 5.75: Transcript copies of *kil* target from equal amounts of total RNA from the batch growth phase samples from a cultivation of *E. coli* JM109-p582 (see 5.6.3).

After an initial low level, there was an increase of about 5-fold in *kil* transcript copies near the end of the exponential phase and this correlated with the idea that stationary phase promoters are activated towards the end of the exponential growth period. After the peak, the concentration seemed to drop to an extent and then increased again during the start of the stationary phase. This could be a case of activation of the P_{fic} promoter initially at the end of exponential phase followed by a global shut down of all systems due to the onset of the stationary phase prior to a reactivation of the promoter once again in the stationary phase. The transcript profile was consistent with the observation by Utsumi *et al.*, (1993) that the P_{fic} promoter is induced by RpoS during the entry into stationary phase.

Transcription of *rpoS*

The *rpoS* target was analysed similarly by means of a RT-qPCR technique but using the primers rpoS-left and rpoS-right (Table 4.3). A standard pJET-rpoS-frag. preparation of 1.8×10^7 copies per reaction was used as a marker to adjust the imported standard curve which is described in section 5.6.5 (under Transcription of *rpoS*). The normalized fluorescence curves for the duplicate samples showed good correlation with each other and the NTC had a high C_T of 30. The melting curve analysis showed normal curves without any non-specific amplicons. Based on the imported standard plot and the observed C_T values, the copies of *rpoS* transcripts present in the RNA from batch-phase samples were calculated and are shown in Fig. 5.76 with the growth curve overlaid.

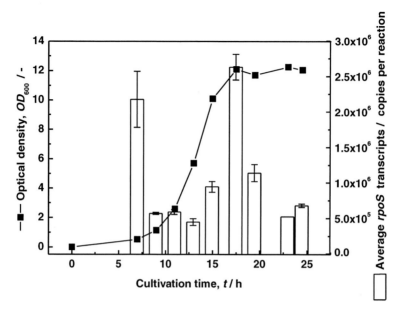

Fig. 5.76: Transcript copies of *rpoS* target from equal amounts of total RNA from the batch growth phase samples from a cultivation of *E. coli* JM109-p582 (see 5.6.3).

The transcript *rpoS* already showed a certain degree of activation during the start of growth but dropped immediately thereafter. The low level of *rpoS* transcript seen in this phase was consistent with the finding that 2 weak promoters of the upstream *nlpD* gene are responsible for low level read-through expression of *rpoS* in growing cells (Lange *et al.*, 1995) and the low level of RpoS protein detected by Tanaka *et al.*, (1993) in exponential phase. At the onset of stationary phase, there was a clear increase in *rpoS* transcript level which corresponded to the anticipation of starvation conditions by the cell and the switching of its metabolism. Studies with transcriptional and translational *lacZ* fusions showed only a nominal change in *rpoS* transcription level during entry into stationary phase in minimal medium and the existence of control at post-transcriptional levels to increase σ^S synthesis rate (Lange & Hennge-Aronis, 1994). The fall of *rpoS* transcript level in this work during the stationary phase could perhaps be explained by the information from literature where Zgurskaya *et al.*, (1997) also found a similar decrease in *rpoS* transcript during stationary phase and reported that RpoS protein concentration increased during starvation as a result of increased protein stability in spite of a reduction in the efficiency of transcription of the gene and translation of transcript. In fact, Lange & Hennge-Aronis (1994) reported a 7-fold increase in σ^S half-life at the onset of starvation in spite of a general protein turnover in this phase which correlates with the 7-fold increase in RpoS protein per cell over the complete growth period from start of exponential phase to stationary phase (Tanaka *et al.*, 1993).

The role of ppGpp in *rpoS* transcription was mentioned in chapter 2.10. The presence of the *relA1* marker on *E. coli* JM109 signifies that in case of stringent response caused by amino acid starvation, only a residual synthetic activity of RelA would be present and therefore accumulation of ppGpp above basal level would not be possible (Metzger *et al.*, 1989). However, since *relA*-independent routes are also possible, *rpoS* expression is not expected to be affected seriously.

Transcription of *rpoD*

For the last target, the primers rpoD-left and rpoD-right were used (Table 4.3) for the RT-qPCR. The standard curve described in section 5.6.5 (under Transcription of *rpoD*) was imported for the analysis. The average transcript copies from duplicate estimations and their standard deviations overlaid with the growth curve are shown in Fig. 5.77.

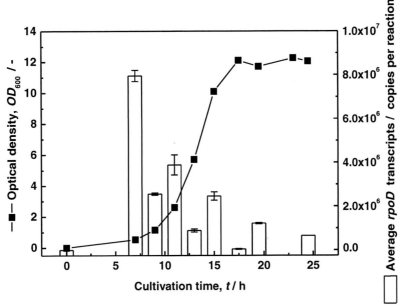

Fig. 5.77: Transcript copies of *rpoD* target from equal amounts of total RNA from the batch growth phase from a cultivation of *E. coli* JM109-p582 (see 5.6.3).

The diagram clearly shows that the *rpoD* transcript levels of this target peaked during the early exponential growth phase before gradually fading away during the stationary phase. This is in accord with the fact that as the primary sigma factor for the transcription of genes during normal growth conditions, the protein subunit corresponding to this transcript is to be present at maximal levels in cells during exponential growth (Keseler *et al.*, 2011). Melting curve analysis of the amplicons from analysis of *rpoD* targets shown in Fig. 5.78 gave a clear distinction in the form of the peak for specific target amplicons and the primer dimers formed in the NTC.

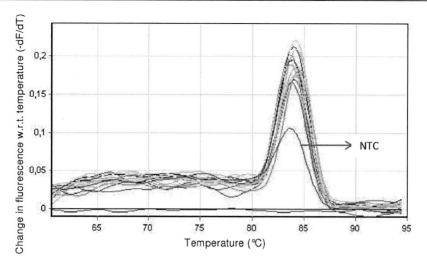

Fig. 5.78: Melting curve analysis of *rpoD* amplification reactions. The denaturation profile for the analysis has been described in chapter 4.8.4.

5.6.5 Chemostat-phase transcript profile for *kil*, *rpoS* and *rpoD*

The next step was the RT-qPCR analysis of the RNA isolated from chemostat-phase samples. Samples were drawn from the chemostat with *E.coli* JM109-p582 from a wide range of space velocities from 0.05 h^{-1}, 0.1 h^{-1}, 0.2 h^{-1}, 0.3 h^{-1} to 0.4 h^{-1} in order to be able to find a significant change in transcript levels of the targets (see 5.6.3). The samples were analysed in duplicate for RT-qPCR in order to derive reliable data. The mean and standard deviations from these quantifications were used to plot the profiles for the amounts of each target. For each target, a set of standards were analysed along with the unknown samples for absolute quantification of transcript copies. The RT-qPCR analysis was performed using the Rotor Gene SYBR Green RT-PCR kit (QIAGEN). The reaction setup was similar to that shown in Table 5.11, with the exception that 80 ng of total RNA was added to each reaction due to availability of higher template RNA concentration. Thus the data for the quantified amounts of the target represent the number of copies of the target present in 80 ng of the total RNA isolated from the cells. The cycling conditions were the same as shown under Table 5.12.

Transcription of *kil*

The normalized fluorescence curves showed good correlation between the duplicate samples except for a few samples which would be discussed later. Standards with a concentration below 7.8×10^3 copies per reaction did not amplify with the same efficiency as other concentrations and hence had to be ignored from the standard curve. Nevertheless, the known standards for absolute quantification extended over 7 orders of magnitude of concentration and contained the threshold cycles for the unknown samples well within their range. The NTC gave an unusually low C_T of 23.6 (Fig. 5.79).

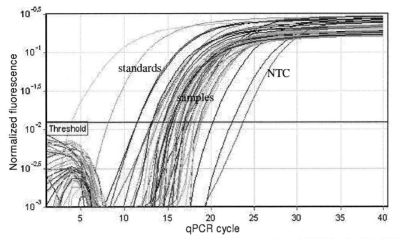

Fig. 5.79: Normalized fluorescence curves for standards, unknown samples and NTC for the RT-qPCR analysis of the *kil* targets of RNA from chemostat-phase samples from a cultivation of *E. coli* JM109-p582.

The standard plot with a correlation coefficient of 0.99 resulted in the following linear correlation.

$$C_T = -2.979 \times \lg(C_r) + 34.375 \qquad (50)$$

Equation (50) was used in the estimation of the *kil* transcript copies shown in Fig. 5.80.

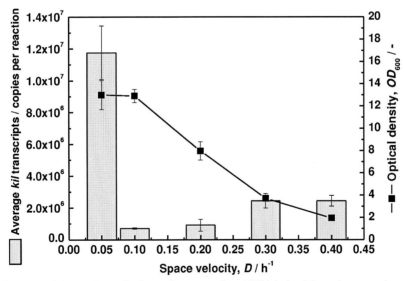

Fig. 5.80: Average *kil* transcript copies in equal amounts of total RNA isolated from chemostat-phase samples over different space velocities, overlaid with the biomass curve from a cultivation of *E. coli* JM109-p582 (see 5.6.3).

The central role of the specific growth rate and its inverse correlation with RpoS protein levels in the exponential phase in batch cultures and in steady-state chemostat cultures was shown with ppGpp as a possible signal linking the two parameters (Ihssen & Egli, 2004). Complementary to this, increased

RpoS levels should be able to directly result in the activation of the P_{fic} promoter and the plot in Fig. 5.80 clearly shows that the *kil* transcript level was highest at the lowest growth rate in the chemostat. This was in accordance with the fact that the P_{fic} promoter is activated during the onset of stationary-phase in a batch cultivation (Miksch *et al.*, 1997a), that corresponds to a low growth rate in a chemostat. A decline of almost 5-fold in the transcript levels towards higher growth rates could also be found.

The abrupt fall in the transcript level at 0.1 h^{-1} in Fig. 5.80 could be related to non-specific amplifications found in the corresponding samples. From the melting curve analysis of the batch-phase samples, a specific peak for the target amplicon at 82.5 °C was to be expected. However, in the melting curve analysis for the chemostat-phase samples shown in Fig. 5.81 all the reactions for the samples from 0.1 h^{-1} had a second peak at a melting temperature of about 5 °C higher than that for the target amplicon. A few of these samples had equal distribution of the expected as well as the second peak whereas a few others had predominantly the expected or the second peak. When these amplified PCR solutions were analysed by agarose gel electrophoresis, an unidentified second band of 400 bp was found. The smaller size of the second peak in Fig. 5.81 could suggest that this amplicon may not have bound a large amount of the dye. In any case, the occurrence of a parallel amplification would have definitely interfered with the target amplification by way of causing a limitation for the ingredients in the PCR reaction. This larger-sized amplicon could have caused a drag on the amplification of the target and skewed its kinetics of amplification while itself amplifying at a very low efficiency. This error could have manifested as a delayed increase in fluorescence and hence the abnormal increase in C_T for these samples from 0.1 h^{-1} and hence the drop seen in Fig. 5.80.

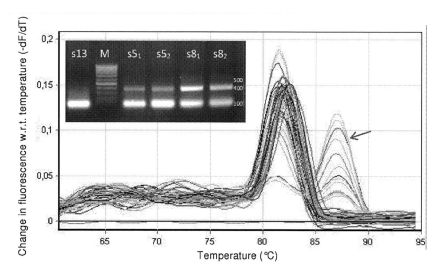

Fig. 5.81: Melting curve analysis of amplicons from RT-qPCR of *kil* targets. Second peaks due to the non-specific amplification of samples from 0.1 h^{-1} are marked with an arrow. (inset) Agarose gel electrophoresis for comparison of a specifically amplified sample (s13) and non-specific amplifications in s5 and s8 (in duplicate). M refers to the 100 bp marker from Plasmid Factory GmbH with band sizes given in bp.

Transcription of *rpoS*

The second target for stationary-phase activation (*rpoS*), was analysed (in duplicate) similar to the previous target, but using the corresponding primers rpoS-left and rpoS-right (Table 4.3). A set of standards (in duplicate) were similarly analysed along with the unknown samples in order to be directly applied in standard curve construction and estimation of the unknown transcript concentrations (Fig. 5.82). These standards ranged from 1.8×10^9 to 1.8×10^3 copies per reaction (C_r). The highest deviations were found in the case of std5 with a given concentration of 1.8×10^6 copies per reaction. Otherwise, the standards clearly followed the expected C_T difference between the logarithms of the concentration and the samples mostly showed good consistency between the duplicates.

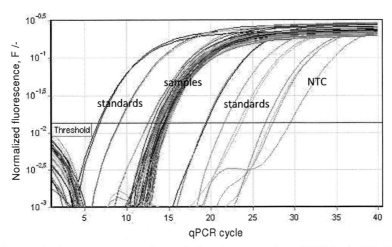

Fig. 5.82: Normalized fluorescence curves for standards, unknown samples and NTC for the RT-qPCR analysis of the *rpoS* targets of RNA from chemostat-phase samples from a cultivation of *E. coli* JM109-p582.

Using the set of standards shown in Fig. 5.82, the following linear equation could be derived.

$$C_T = -3.478 \times \lg(C_r) + 37.829 \tag{51}$$

Based on the value of the slope (-3.478) an efficiency of amplification of 0.938 could be calculated. The calculation of the *rpoS* transcript copies in the unknown samples based on the standard equation revealed a pattern that was to be expected for a basal level during high growth rates and activation at low growth rates (Fig. 5.83). However, an unusual increase in the transcript copies was found at the highest space velocity of 0.4 h^{-1}. This was also seen in the *rpoS* transcript profile during exponential growth in batch phase (Fig. 5.76). The overall change in transcript copies over the range of specific growth rates was only a factor of 2.3.

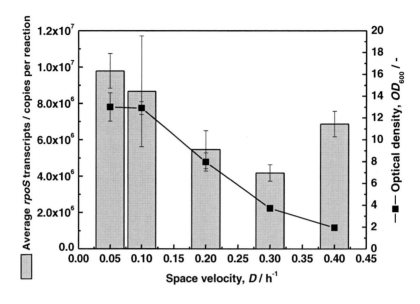

Fig. 5.83: Average *rpoS* transcript copies in equal amounts of total RNA isolated from chemostat-phase samples over different space velocities, overlaid with the biomass curve from a cultivation of *E. coli* JM109-p582 (see 5.6.3).

At the protein level, Teich *et al.*, (1999) showed the ability of the cell to dynamically shift σ^S protein concentration from the baseline within a short time frame as a response to step-wise change in dilution rate in a glucose-limited chemostat. A peaking at 15 min after dilution range change was followed by falling back to base level. This reinforces the argument by Lange & Hengge-Aronis (1994) that the cellular level of σ^S is dynamically controlled at multiple levels allowing the cells to adapt swiftly to varying nutrient levels. With a computer-controlled gradual reduction of dilution rate Teich *et al.*, (1999) found a small increase in σ^S protein and significant increase of ppGpp levels only at very low growth rates of 0.1 h^{-1}. This may explain why the *kil* transcript concentration was seen to be increased substantially only in the lowest space velocity (Fig. 5.80). The other space velocities applied would have caused changes in *rpoS* transcript level and momentary changes in RpoS protein level too, but the steady state RpoS protein levels were maintained again at the baseline level as the cells adjust to the new conditions within a couple of hours.

Similar to the batch results, some *rpoS* transcription should be possible from the *nlpD* promoter as a bicistronic mRNA transcript during growth at high specific growth rates (Lange *et al.*, 1995) and this could have led to the relatively high basal level of transcript seen at space velocities of 0.3 h^{-1}-0.4 h^{-1} in Fig. 5.83. RpoS expression is also controlled at the level of protein translation and post-translation (Landini *et al.*, 2014). Post-translational control includes protein degradation and binding control to RNAP core enzyme. So the *rpoS* transcript profile measured here cannot be expected to give a similar *kil* profile. This is because P$_{fic}$ is dependent on RpoS protein and not *rpoS* mRNA. Taken together, one may not see a distinct increase in *rpoS* transcript level at low growth rates, since the RpoS protein level may rather be increased by reduced proteolytic degradation (Becker *et al.*, 2000).

Transcription of *rpoD*

The RT-qPCR analysis of the *rpoD* targets from the chemostat-phase samples was performed along with standard pJET-rpoD-frag. plasmid preparations ranging over 8 orders of magnitude of concentration ranging from 7.02×10^9 to 7.02×10^2 copies per reaction (C_r). The standard curve resulted in the following correlation which was used for the estimation of the transcript copies in unknown samples.

$$C_T = -3.235 \times \lg(C_r) + 35.935 \tag{52}$$

The huge margin of error in the samples from high growth rates notwithstanding, the transcript profile for *rpoD* correlated with the expectation of high activation during the phase of maximum growth rate and deregulation at lower growth rates corresponding to starvation (Fig. 5.84). This profile served as a good comparison for the profiles seen earlier for *kil* and *rpoS*. The factor of change in transcript copies in this case was only 1.73.

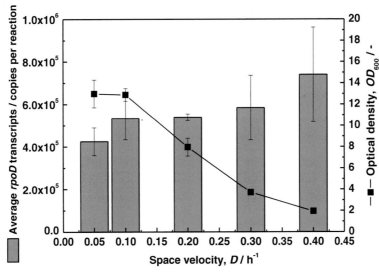

Fig. 5.84: Average *rpoD* transcript copies in equal amounts of total RNA isolated from chemostat-phase samples over different space velocities, overlaid with the biomass curve from a cultivation of *E. coli* JM109-p582 (see 5.6.3).

In order to derive a complete picture of the nature of change in the transcript levels during the various space velocities in the chemostat, all 3 transcripts are shown together in Fig. 5.85. To conclude, as beautifully elucidated in Herbert *et al.*, (1956), attaining a steady-state in a chemostat is not a great challenge since it is the only natural stable configuration for the culture and which in their words is an "inevitable" outcome. The challenge though is to maintain the steady-state under carefully controlled conditions especially for experiments that aim to analyse changes in the mRNA populations in the cells.

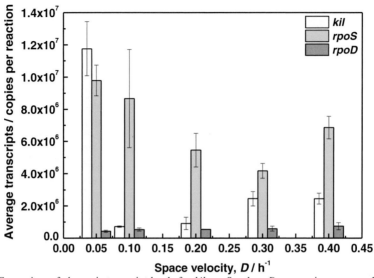

Fig. 5.85: Comparison of change in transcript levels for *kil*, *rpoS* and *rpoD* over various space velocities in a chemostat of *E. coli* JM109-p582 (see 5.6.3).

It must be emphasized though, that the activity of the promoter P_{fic} to synthesize *kil* transcripts is only a part of a more elaborate process. The actual extracellular secretion of target protein brought about by the *kil* gene product BRP still depends on effective translation, folding, modification and transport of the BRP through the cell membrane. Although the use of plasmid DNA standards for the quantification of cytokine mRNAs from rabbit spleen cells has been reported (Godornes *et al.*, 2007), absolute quantification of mRNA transcripts could be greatly improved by using RNA standards and performing a RT-qPCR for the standard analysis similar to the samples. This would take into account the errors during the reverse transcription step which were missed when using DNA standards. The pJET-fragment plasmids created in this work could be directly used for *in vitro* transcription from the T7 promoter of pJET to generate the respective RNA transcripts which could then be analysed by spectrophotometry and used as standard copies.

5.7 Analysis of relative plasmid abundance

Although plasmids are regarded as autonomous molecules, the native regulatory mechanisms of a plasmid such as its replication control are subject to host-cell growth-rate dependent effects (Klumpp, 2011). Measurement of the relative plasmid abundance in the cells gives an approximation of the average plasmid copy number since normally a distribution of plasmid copy numbers occurs among the cells in a population (Friehs, 2004). Additionally, cells with higher than average copy number may not necessarily result in high specific productivity due to limitations in the host protein synthesis machinery. Changes in relative plasmid abundance over various growth rates are also important to track in order to account for this factor while interpreting the results of *kil* transcript concentration changes from the RT-qPCR experiments. A separate chemostat with the strain *E. coli* JM109-p582 with kanamycin selection pressure was carried out to study the effects of the growth rate on plasmid content in the cells. The cultivation was carried out under the same conditions as described for the RT-qPCR analysis in chapter 5.6.3.

5.7.1 Standard curve for *bla*

Since the beta-lactamase gene (ampicillin resistance) is present as a single copy on the plasmid p582, a target region within the sequence of this gene acted as an optimal complement to the single copy genomic gene *rpoS*, in order to determine the ratio of the amounts of each of the targets. Serial dilutions with known concentrations of the plasmid p582 were analysed in a qPCR experiment (Fig. 5.86) using the bla-right and bla-left primers (Table 4.3).

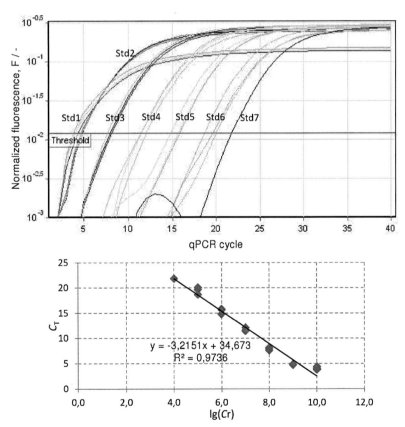

Fig. 5.86: (above) qPCR analysis of known amounts of p582 plasmid templates using the primers bla-left and bla-right for the amplification of a target region in the beta-lactamase gene. (below) Standard plot in the form of C_T versus logarithm of concentration of target in copies per reaction.

The standards ranged in concentration from 1.0×10^{10} (std1) to 1.0×10^{4} (std7) copies per reaction, each measured in triplicate. Except the last two reactions of std7, all others amplified according to the expected C_T values and showed reasonable consistency. Std1, due to the highest concentration of template, showed signs of inhibition which resulted in a higher C_T than the expected value. Based on the standard curve, the efficiency of amplification (E_x) was found to be 1.05 and the slope of the standard curve (m) was -3.215. For *rpoS*, the standard curve described in chapter 5.6.5 (under Transcription of *rpoS*) was used for the estimation.

5.7.2 Batch phase

The procedure for the measurement of relative plasmid abundance has been described in chapter 4.9. The C_T for the *bla* targets were consistently lower than that for the *rpoS* targets which was a sign that the assay was working normally. The concentration of the targets were calculated on the basis of the respective standard correlations applied to the C_T values of unknown samples (Fig. 5.87). Although a slight increase could be seen towards the end of the batch fermentation, overall, the trend seen for the values of relative plasmid abundance did not allow inferring any strong conclusions. There was an initial drop seen in the mean relative plasmid abundance at the beginning of the exponential growth phase followed by a period of stability and then finally again a slight increase during mid-stationary phase. One of the amplification curves for *bla* for the sample 6 h batch time was extremely irregular and had to be omitted and therefore no mean calculation was possible for this point.

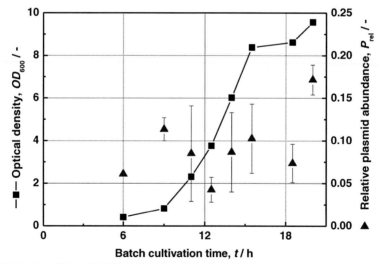

Fig. 5.87: Cultivation of *E. coli* JM109-p582 and analysis of relative plasmid abundance over the batch growth phase, according to the procedure detailed in chapter 4.9.

5.7.3 Chemostat phase

Samples were drawn at 4 different time points at each of the space velocities from 0.1 h^{-1} to 0.4 h^{-1} from the chemostat cultivation of *E. coli* JM109-p582 described under chapter 5.7. The average relative plasmid abundance for these space velocities representing different specific growth rates of the strain, are shown in Fig. 5.88. Due to the different threshold values set for this analysis, the values for P_{rel} are distinctly higher than those seen for the batch-phase in Fig. 5.87.

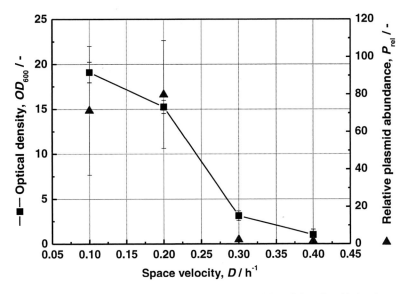

Fig. 5.88: Chemostat cultivation of *E. coli* JM109-p582 and analysis of the relative plasmid abundance over various specific growth rates, according to the procedure detailed in chapter 4.9.

The data shown in Fig. 5.88 suggested the existence of an optimal growth rate for maximum plasmid production in the cells which in this case was a relative plasmid abundance of 79 at a space velocity of 0.2 h^{-1}. Over higher growth rates, the relative plasmid abundance fell and maintained stable but the measurements were affected by the very low cell concentrations for these samples. The data correlated at least partially with the profile seen during batch growth in that, the plasmid content was lowest during exponential growth rates while being higher before and after this phase. However, the large error margins in the estimations need to be considered.

Replication of ColE1-type plasmids is initiated by the primer RNA II which could be regulated by binding of a short-lived antisense RNA (RNA I). In the absence of binding by RNA I ahead of the replication origin, the primer can mature and continue to replicate the plasmid DNA. The *rop* gene product (Rom) promotes the binding of RNA I to RNA II and increases the sensitivity of the control (Klumpp, 2011). In the pUC–group of plasmids, the deletion of the *rop* gene and a point mutation in the RNA II transcript together relieve the plasmid from control in such a way that the copy number is increased to about 70 per cell at 37 °C (Lin-Chao *et al.*, 1992). Data summarized from the literature for plasmid pBR322 indicates an intricate connection between cell growth rate and plasmid replication control and a slight increase in plasmid copy number per cell at high growth rates (Klumpp, 2011). The plasmid p582 is based on the pUC19 origin, and there are multiple factors that determine the average plasmid copy number in a cell. Due to this, and the fact that the method employed here anyway gives only an estimate of plasmid-borne sequence relative to a genome-borne sequence, the number in itself may not correspond to the plasmid copy number and, therefore, it is referred to here as relative plasmid abundance in the cells. This term was also used in other reports in the literature (Dong et al., 2010; Goh & Good, 2008.)

The contrasting observations in the literature with regard to the relation between growth rate and plasmid DNA content in the cells has been summarized by Lara & Ramírez (2012) with the conclusion that on the basis of several observations, a negative correlation is the general trend to consider. However, the observation in Fig. 5.88 that on the whole the data pointed to a maximum in relative plasmid abundance at an optimal space velocity of 0.2 h^{-1} also finds support from various reports. Continuous culture with E. coli strain HB101 carrying a pMB1-derivative in a M9-supplemented medium showed strong evidence for the possibility of an optimal specific growth rate for which the plasmid content per cell reached a maximum and correlated with observations for batch experiments (Seo & Bailey, 1986). The same strain with either pBR322 or a large molecular mass derivative thereof, was grown in a 500 mL chemostat culture in complex medium and a similar maximum for the plasmid copy number per genomic DNA was found to exist at an optimal dilution rate of about 1 h^{-1}. However, due to the nature of the medium used, this result may have to be interpreted with caution (Reinikainen & Virkajärvi, 1989).

The measurement of relative plasmid abundance by qPCR could definitely be improved by isolating total DNA from the bacterial cells using a commercial kit rather than a simple thermal lysis method. The latter method was followed here, in order to develop a quick procedure and to have a uniform treatment. The idea had been to avoid the added steps for total DNA isolation which may have caused more chances of variation in template DNA amounts among the samples due to different efficiencies of isolation. The relative plasmid abundance data were relevant in order to be taken together with the RT-qPCR data of target transcripts in the cells, so that the role of plasmid copy number changes at different growth rates in affecting the amounts of transcript copies could be accounted for. Therefore, the increased levels of kil transcripts at low specific growth rates seen in Fig. 5.80 may have partly been aided by the increased availability of the template due to the higher relative plasmid abundance at this specific growth rate in comparison to the highest specific growth rate.

5.8 Analysis of P_{fic} activity by GFP fluorescence measurement

The gfp gene has earlier proved to be useful in monitoring stress responses in non-growing cells of E. coli JM105 by fusing it to promoters of stress-related genes such as those for σ^{32}, ClpB and DnaK and testing various kind of stresses including heat shock and ethanol addition (Cha et al., 1999). The compatibility of this reporter protein to automated and online measurement and its non-invasive nature are advantages over the traditional β-galactosidase based assay, as shown in an earlier work for screening a library of synthetic stationary-phase promoters (Miksch et al., 2006). Therefore, the second method for analysis of promoter activity employed in this work was the use of an unstable green fluorescent protein variant as a reporter for activation of the P_{fic} promoter under various specific growth rates.

5.8.1 Inverse PCR

The plasmid pET-24a(+)-gfp+-lva (Table 4.2) was used as a template for inverse PCR with primers 41 and 42 (Table 4.3). The target amplicon would be 4037 bp long. The primers contained the P_{fic} promoter in their 5'-overhangs and were designed to replace the T7 promoter on the template plasmid. Simultaneously, the primers were designed to bind in a region so that the *lacI* gene could also be deleted, thereby reducing the plasmid size. The primers contained challenging T_m characteristics such as 61_1, 76_2 and 63_1, 81_2 for primers 41 and 42 respectively. The subscript 1 referred to the first cycle when the overhangs could not bind to any template and subscript 2 to the subsequent cycles when the primers bound along the entirety of their length. The reaction setup and cycling conditions for optimization through gradient PCR are shown in Tables 5.13 and 5.14.

Table 5.13: Reaction set up for inverse PCR for replacing P_{T7} with P_{fic} upstream to *gfp+*-lva on pET-24a(+)-gfp+-lva.

Component	Volume (μL)
Template plasmid (67 ng μL^{-1})	0.5
Phusion GC buffer	10
10 mM dNTPs	1
100 μM Primer 41	0.25
100 μM Primer 42	0.25
Phusion Hot-Start polymerase	0.5
Water	37.5
Total	**50**

Table 5.14: Cycling protocol for inverse PCR for replacing P_{T7} with P_{fic} upstream to *gfp+*-lva on pET-24a(+)-gfp+-lva.

Step		Temp (°C)	Time (s)
Initial denaturation		98	30
Cycling 10x	Denaturation	98	10
	Annealing gradient	59 → 63	20
	Extension	72	82
Cycling 25x	Denaturation	98	10
	Combined annealing + extension	72	90
Final extension		72	300

Testing a fraction of the reaction on agarose gel showed that the gradient PCR yielded the amplicon of expected size. The reactions were pooled together and DpnI-treated to lyse the template DNA. The linear DNA product was electrophoresed on a gel, purified, phosphorylated at the 5'- ends using T4 polynucleotidekinase, followed by self-ligation using T4 DNA ligase. The plasmids from four of the resulting clones were sequenced and verified to contain the P_{fic} promoter correctly ahead of the *gfp+-lva* gene in three of them (clones 1, 3 and 4). This construct was designated pRS-ficGFP and its schematic map is shown in Fig. 8.2 (Appendix).

The three verified clones of pRS-ficGFP served as 3 replicates and were transformed into the control strain *E. coli* JM109. Identities of the plasmids could be verified by single cut restriction using the enzyme SacI. In an initial shake-flask cultivation study, the growth and *gfp* expression profiles of the 3 clones were tested in SGA medium at 37 °C (Fig. 5.89). The fluorescence normalized to the optical density, showed a clear increase, albeit very low, at the end of the exponential growth phase.

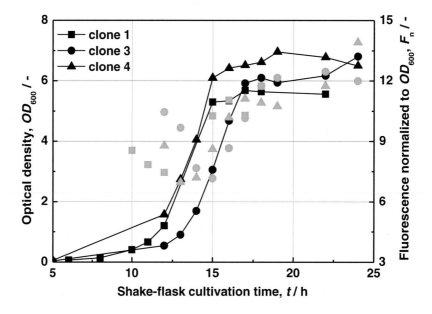

Fig. 5.89: Shake-flask cultivation of 3 clones of *E. coli* JM109-pRS-ficGFP to test growth and GFP+-LVA expression to analyse the promoter P_{fic} activity. Each data point for optical density representing growth of the culture (black) is the average from duplicate measurements. The data points for normalized fluorescence (green) are the average from triplicate measurements. Shake-flask cultivation was according to the general procedures described in chapter 4.6.1.

5.8.2 Batch-phase analysis of variation in promoter activity

The clone 1 from Fig. 5.89 was studied in batch mode (1 L) in the in-house fermenter using the same medium (SGA supplemented with thiamine) as for the shake-flask cultivation and the process was carried out at 37 °C. Figure 5.90 with the results of this batch process shows that the activation of the promoter did take place after the end of the exponential growth phase. The very low increase seen in the normalized fluorescence values could be the result of the fact that P_{fic} is classified as a weak promoter.

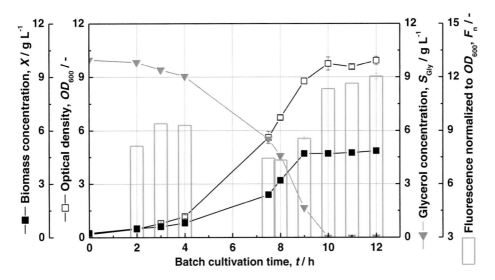

Fig. 5.90: Batch cultivation of clone 1 of *E. coli* JM109-pRS-ficGFP for analysis of the P_{fic} promoter activity by way of measurement of GFP+-LVA fluorescence over the batch growth phase. The batch process was carried out according to the descriptions for cultivation in the in-house fermenter given in chapter 4.6.2.

5.8.3 Chemostat-phase analysis of variation in promoter activity

The strain *E. coli* JM109-pRS-ficGFP described in the batch process in chapter 5.8.2 was studied under continuous cultivation in SGA medium at 37 °C in order to make final analyses on the promoter activity under various specific growth rate conditions. The results shown in Fig. 5.91 reflected the data seen from RT-qPCR for the activity of the P_{fic} promoter in that the lowest growth rate represented by the space velocity of 0.1 h^{-1} showed the maximum normalized GFP fluorescence. This was in line with the expectation of the P_{fic} promoter to be activated under starvation conditions. The biomass growth however, showed characteristics of first-order kinetics with the linear increase of steady state substrate concentration particularly pointing to this behaviour. Significantly, similar to the *kil* transcript profile seen from the RT-qPCR experiments in Fig. 5.80, the activity of the P_{fic} promoter seemed to decrease from its maximum value at the lowest space velocity to a far lower level during the intermediate growth rates and later stabilising during the maximum space velocities of 0.3 h^{-1} and 0.4 h^{-1}. These data offer useful insights into the regulation of the P_{fic} promoter which was used for the control of the *kil* gene responsible for extracellular release of recombinant proteins in plasmid p582 and its derivatives.

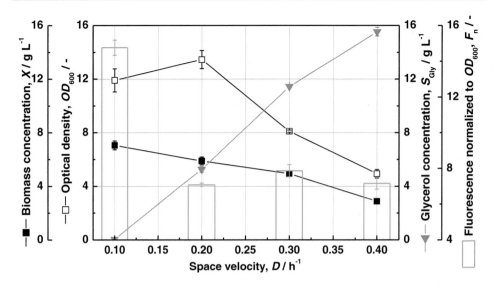

Fig. 5.91: Chemostat cultivation of clone 1 of *E. coli* JM109- pRS-ficGFP for analysis of the P$_{fic}$ promoter activity by means of measurement of normalized GFP+-LVA fluorescence over various specific growth rates. The cultivation was carried out in the in-house fermenter described in chapter 4.6.2 under conditions described in chapter 4.6.3.

6 Conclusion

In this work, the reference system *E. coli* JM109-p582 was demonstrated to be able to stably express recombinant β-glucanase in the extracellular medium in a continuous cultivation with antibiotic selection pressure. It was shown for first time that a continuous operation of this recombinant bioprocess was possible even though the outer membrane of the cells were continuously under potential damage by the action of the BRP. A volumetric β-glucanase productivity of 128.6 kU L^{-1} h^{-1} represented an enormous potential for this continuous process. The product expression in the extracellular medium and the related values of volumetric productivity represent true steady values with respect to long-term operation of the chemostat and differ particularly from strategies that require a distinct point of induction and then measure product on the basis of time post-induction (Vaiphei *et al.*, 2009). For further development of this system, a constitutive promoter for *kil* could be tested to see if it is possible to operate in a similarly stable fashion in continuous cultivation. If so, it could allow operation of higher flow rates since the promoter for product transport from the cells would still be active. This may allow higher productivity if the biomass would not be seriously challenged by the damage to the outer membranes. Despite its several advantages, the problems with long-term stability were pointed out to be one of the serious impedances to the application of continuous culture for industrial production (Fu *et al.*, 1993). The antibiotic-free auxotrophic complementation strategies shown in this doctoral thesis are examples of anabolism-based (leucine) and catabolism-based (glycerol) complementation systems, and represent competitive and sustainable solutions to the problem of maintaining plasmid stability in long-term cultivation processes.

The establishment of the auxotrophic system *E. coli* JW5807-2-pFC2 based on *leuB* complementation allowed stable antibiotic-free cultivation, but at the cost of productivity. The auxotrophic system showed lower extracellular activity (509 kU L^{-1}) in a batch process and a lower volumetric productivity of 24.9 kU L^{-1} h^{-1} in a continuous process. The decrease in isolated plasmid content was found to be up to 50% in comparison to 90% in the control strain in the absence of antibiotic selection pressure. The segregational plasmid stability could be maintained at 100% throughout without any antibiotics. Efforts at a two-stage chemostat process offered very interesting information on the kinetics of biomass growth and product expression in a cascade system but resulted in no significant improvement of the productivities. The deletion of the kanamycin resistance gene from the construct pFC2 marked one step towards realizing a completely antibiotic/ antibiotic resistance gene- free process, though the success could not be extended to selection of clones containing a deletion of both ampicillin and kanamycin resistance genes. Using effective strategies like the chew-back and anneal method (Gibson *et al.*, 2009), this step could be pursued further along with the removal of large portions of the plasmid that are unnecessary, thus drastically reducing the plasmid size.

The establishment of the *E. coli* JW3890-2-pFC4 system proved the possibility to use the crucial metabolic link TpiA as an effective selection principle. The advantage here was that a moderate amount of growth was possible in complex medium, but in minimal medium, the plasmid-carrying cells alone could grow. However, the losses to productivity were even greater here: the maximum activity in a batch process was only 132 kU L^{-1} with a possible limitation in extracellular secretion from the periplasm and the maximum volumetric productivity was 6.5-10 kU L^{-1} h^{-1} in continuous cultivation. Similar to the *leuB* system, the *tpiA* complementation system too showed high plasmid stability in the absence of antibiotics, thus proving it to be an effective selection principle. Attempts at moderating the expression of the *tpiA* complementation gene by using weak promoters CP19 and CP33 resulted in better biomass production and a 5-fold increase in β-glucanase expression in one CP33 clone. Deletion of the target *tpiA* gene in the original reference strain *E. coli* JM109 allowed the freedom from the use of Keio strains and also offered a stable gene deletion locus with reduced risk of recombination with the plasmid-borne *tpiA* gene. However, efforts at reproducing the efficiency of the reference system (JM109-p582) were not successful and this probably represented an inevitable cost of using auxotrophic complementation. The metabolic costs for maintaining high-copy recombinant plasmids in order to survive in SGA minimal medium probably resulted in an inevitable loss of efficiency of target gene expression and secretion and thus reduced volumetric productivity. One common feature with the auxotrophic strains has been the poor growth characteristics. Therefore, another direction to look at for development of this continuous cultivation system would be rational medium design for better growth of the strain.

Although in general, both the auxotrophic complementation systems (*leuB* and *tpiA*) could not reach the extracellular recombinant β-glucanase productivity achieved with the reference strain JM109-p582, the latter had been possible only with antibiotic addition. Thus, the most important value addition brought by the auxotrophic complementation systems for long-term continuous cultivation and recombinant product expression has been the possibility of a sustainable and stable antibiotic-free process, with possibilities to improve to a completely antibiotic/antibiotic resistance gene-free system.

The analysis of the transcripts of *kil*, *rpoS* and *rpoD* by RT-qPCR gave for the first time a direct comparison of the activities of P$_{fic}$ and the promoter of *rpoS* and *rpoD* in samples from different specific growth rates. The profiles of these transcripts were in accordance with the expectations based on literature that the P$_{fic}$ is dependent on RpoS for expression and is activated during the stationary phase of batch which corresponded to the space velocities for low specific growth rate in the chemostat. This offered insight into the control of *kil* gene activation at the molecular level and these findings were supported by observations from transcriptional fusions using an analytic method based on GFP fluorescence. The analysis of relative plasmid abundance levels as a function of specific growth rate pointed to an optimal space velocity that resulted in maximum plasmid copy per genome equivalent. Taking these factors together, an optimal operational space velocity can be determined that results in the best compromise between biomass, recombinant gene expression, *kil* gene activation and extracellular recombinant enzyme productivity. By replacing the model protein β-glucanase with other industrially relevant enzymes such as lipases, proteases or phytases, significant advances in the application of continuous culture in white biotechnology are possible.

7 References

Achmüller, C., Kaar, W., Ahrer, K., Wechner, P., Hahn, R., Werther, F., Schmidinger, H., Cserjan-Puschmann, M., Clementschitsch, F., Striedner, G., Bayer, K., Jungbauer, A., & Auer, B. (2007). N^{pro} fusion technology to produce proteins with authentic N termini in *E. coli. Nat Methods, 4* (12), 1037-1043.

Ahn, W. S., Jeon, J-J., Jeong, Y-R., Lee, S. J., & Yoon, S. K. (2008). Effect of culture temperature on erythropoietin production and glycosylation in a perfusion culture of recombinant CHO cells. *Biotechnol Bioeng, 101* (6), 1234–1244.

Albano, C. R., Randers-Eichhon, L., Chang, Q., Bentley, W. E., & Rao, G. (1996). Quantitative measurement of green fluorescent protein expression. *Biotechnol Tech, 10* (12), 953–958.

Alexander, P., Oriel, P. J., Glassner, D. A., & Grulke, E. A. (1989). Continuous culture of *Escherichia coli* for extracellular production of recombinant amylase. *Biotechnol Lett, II* (9), 609–614.

Andersen, J. B., Sternberg, C., Poulsen, L. K., Bjørn, S. P., Givskov, M., & Molin, S. (1998). New unstable variants of green fluorescent protein for studies of transient gene expression in bacteria. *Appl Environ Microbiol, 64* (6), 2240–2246.

Aono, R. (1989). Release of penicillinase by *Escherichia coli* HB101 (pEAP31) accompanying the simultaneous release of outer-membrane components by Kil peptide. *Biochem J, 263* (1), 65-71.

Baba, T., Ara, T., Hasegawa, M., Takai, Y., Okumura, Y., Baba, M., Datsenko, K. A., Tomita, M., Wanner, B. L., & Mori, H. (2006). Construction of *Escherichia coli* K-12 in-frame, single-gene knockout mutants: the Keio collection. *Mol Syst Biol, 2*: 2006.0008. doi:10.1038/msb4100050.

Bakker, W. A. M., Schäfer, T., Beeftink, H. H., Tramper, J., & De Gooijer, C. D. (1996). Hybridomas in a bioreactor cascade: modeling and determination of growth and death kinetics. *Cytotechnology, 21* (3), 263–277. doi:10.1007/BF00365349.

Baneyx, F. (1999). Recombinant protein expression in *Escherichia coli. Curr Opin Biotechnol, 10* (5), 411–421.

Becker, G., Klauck, E., & Hengge-Aronis, R. (2000). The response regulator RssB, a recognition factor for σ^S proteolysis in *Escherichia coli*, can act like an anti-σ^S factor. *Mol Microbiol, 35* (3), 657–666.

Bertani, G. (1951). Studies on lysogenesis. I. The mode of phage liberation by lysogenic *Escherichia coli. J Bacteriol, 62* (3), 293–300.

Beshay, U., Miksch, G., & Flaschel, E. (2007a). Improvement of a *β*-glucanase activity test by taking into account the batch reactor balance of the test system. *Bioproc Biosys Eng, 30* (4), 251–259.

Beshay, U., Miksch, G., Friehs, K., & Flaschel, E. (2007b). Improved β-Glucanase production by a recombinant *Escherichia coli* strain using zinc-ion supplemented medium. *Eng Life Sci, 7* (3), 253–258. doi:10.1002/elsc.200620191

Beshay, U., Miksch, G., Friehs, K., & Flaschel, E. (2007c). Increasing the secretion ability of the *kil* gene for recombinant proteins in *Escherichia coli* by using a strong stationary-phase promoter. *Biotechnol Lett, 29* (12), 1893-1901.

Beshay, U., Miksch, G., Friehs, K., & Flaschel, E. (2009). Integrated bioprocess for the production and purification of recombinant proteins by affinity chromatography in *Escherichia coli*. *Bioproc Biosys Eng*, *32* (2), 149–158. doi:10.1007/s00449-008-0227-3

Bill, R. M. (2014). Playing catch-up with *Escherichia coli*: using yeast to increase success rates in recombinant protein production experiments. *Front Microbiol*, *5*: 85. doi:10.3389/fmicb.2014.00085

Binnewitt, S. V. (2013). *Verkleinerung eines rekombinanten Auxotrophie-Komplementationsplasmids mittels der Entfernung von Antibiotikaresistenzgenen*, Bachelor thesis, Faculty of Technology, Bielefeld University.

Blanch, H. W., & Clark, D. S. (1996). *Biochemical Engineering*. Marcel Dekker, Inc. New York, USA.

Borriss, R., Olsen, O., Thomsen, K. K., & von Wettstein, D. (1989). Hybrid bacillus endo-(1-3,1-4)-beta-glucanases: construction of recombinant genes and molecular properties of the gene products. *Carlsberg Res Commun*, *54* (2), 41-54.

Borsuk, S., Mendum, T. A, Fagundes, M. Q., Michelon, M., Cunha, C. W., McFadden, J., & Dellagostin, O. A. (2007). Auxotrophic complementation as a selectable marker for stable expression of foreign antigens in *Mycobacterium bovis* BCG. *Tuberculosis*, *87* (6), 474-480.

Boyce, A, & Walsh, G. (2007). Production, purification and application-relevant characterisation of an endo-1,3(4)-β-glucanase from *Rhizomucor miehei*. *Appl Microbiol Biotechnol*, *76* (4), 835-841.

Bull, A. T. (2010). The renaissance of continuous culture in the post-genomics age. *J Ind Microbiol Biotechnol*, *37* (10), 993–1021.

Bustin, S. A. (2000). Absolute quantification of mRNA using real-time reverse transcription polymerase chain reaction assays. *J Mol Endocrinol*, *25* (2), 169–193.

Bustin, S. A, Benes, V., Garson, J. A, Hellemans, J., Huggett, J., Kubista, M., Mueller, R., Nolan, T., Pfaffl, M. W., Shipley, G. L., Vandesompele, J., Wittwer, C. T. (2009). The MIQE guidelines: minimum information for publication of quantitative real-time PCR experiments. *Clin Chem*, *55* (4), 611–622.

Çakar, Z. P., Sauer, U., & Bailey, J. E. (1999). Metabolic engineering of yeast: the perils of auxotrophic hosts. *Biotechnol Lett*, *21* (7), 611–616.

Cascales, E., Buchanan, S. K., Duché, D., Kleanthous, C., Lloubès, R., Postle, K., Riley, M., Slatin, S., & Cavard, D. (2007). Colicin biology. *Microbiol Mol Biol Rev*, *71* (1), 158–229. doi:10.1128/MMBR.00036-06

Cavard, D. (1991). Synthesis and functioning of the colicin E1 lysis protein: comparison with the colicin A lysis protein. *J Bacteriol*, *173* (1), 191–196.

Celestino, K. R. S., Cunha, R. B., & Felix, C. R. (2006). Characterization of a β-glucanase produced by *Rhizopus microsporus* var. *microsporus*, and its potential for application in the brewing industry. *BMC Biochem*, *7*: 23. doi:10.1186/1471-2091-7-23

Cha, H. J., Srivastava, R., Vakharia, V. N., Rao, G., & Bentley, W. E. (1999). Green fluorescent protein as a noninvasive stress probe in resting *Escherichia coli* cells. *Appl Environ Microbiol*, *65* (2), 409–414.

Choct, M. (2006). Enzymes for the feed industry: past, present and future. *World Poultry Sci J, 62* (1), 5-15.

Choi, J. H., & Lee, S. Y. (2004). Secretory and extracellular production of recombinant proteins using *Escherichia coli*. *Appl Microbiol Biotechnol, 64* (5), 625–635. doi:10.1007/s00253-004-1559-9.

Cocaign-Bousquet, M., Guyonvarch, A., & Lindley, N. D. (1996). Growth rate-dependent modulation of carbon flux through central metabolism and the kinetic consequences for glucose-limited chemostat cultures of *Corynebacterium glutamicum*. *Appl Environ Microbiol, 62* (2), 429–436.

Cranenburgh, R. M., Lewis, K. S., & Hanak, J. A. J. (2004). Effect of plasmid copy number and *lac* operator sequence on antibiotic-free plasmid selection by operator-repressor titration in *Escherichia coli*. *J Mol Microbiol Biotechnol, 7* (4), 197-203.

Datsenko, K. A, & Wanner, B. L. (2000). One-step inactivation of chromosomal genes in *Escherichia coli* K-12 using PCR products. *Proc Natl Acad Sci USA, 97* (12), 6640-6645.

Davis, R. H., Ramirez, W. F., & Chatterjee, A. (1990). Optimal chemostat cascades for periplasmic protein production. *Biotechnol Prog, 6* (6), 430–436. doi:10.1021/bp00006a005.

De Marco, A. (2009). Strategies for successful recombinant expression of disulfide bond-dependent proteins in *Escherichia coli*. *Microb Cell Fact, 8*: 26. doi:10.1186/1475-2859-8-26

Degryse, E. (1991). Stability of a host-vector system based on complementation of an essential gene in *Escherichia coli*. *J Biotechnol, 18* (1-2), 29–40.

Dekker, N., Tommassen, J., & Verheij, H. M. (1999). Bacteriocin release protein triggers dimerization of outer membrane phospholipase A in vivo. *J Bacteriol, 181* (10), 3281–3283.

Derveaux, S., Vandesompele, J., & Hellemans, J. (2010). How to do successful gene expression analysis using real-time PCR. *Methods, 50* (4), 227–230. doi:10.1016/j.ymeth.2009.11.001

Divate, R., Menon, V., & Rao, M. (2013). Approach towards biocatalytic valorisation of barley β-glucan for bioethanol production using 1,3-1,4 β-glucanase and thermotolerant yeast. *Int Biodeter Biodegr, 82*, 81–86.

Dong, W-R., Xiang, L-X., & Shao, J-Z. (2010). Novel antibiotic-free plasmid selection system based on complementation of host auxotrophy in the NAD *de novo* synthesis pathway. *Appl Environ Microbiol, 76* (7), 2295–2303.

Dragan, A. I., Pavlovic, R., McGivney, J. B., Casas-Finet, J. R., Bishop, E. S., Strouse, R. J., Schenerman, M. A., Geddes, C. D. (2012). SYBR Green I: fluorescence properties and interaction with DNA. *J Fluoresc, 22* (4), 1189–1199. doi:10.1007/s10895-012-1059-8

Dunn, I. J., Heinzle, E., Ingham, J., & Přenosil, J. E. (2003). *Biological Reaction Engineering* (2nd edition). Weinheim: WILEY-VCH Verlag GmbH & Co. KGaA.

Durany, O., Bassett, P., Weiss, A. M. E., Cranenburgh, R. M., Ferrer, P., López-Santín, J., de Mas, C., & Hanak, J. A. J. (2005). Production of fuculose-1-phosphate aldolase using operator-repressor titration for plasmid maintenance in high cell density *Escherichia coli* fermentations. *Biotechnol Bioeng, 91* (4), 460-467.

Eckhardt, T. (1978). A rapid method for the identification of plasmid desoxyribonucleic acid in bacteria. *Plasmid, 1* (4), 584–588.

Enfors, S., & Häggström, L. (2000). *Bioprocess Technology Fundamentals and Applications.* Stockholm: Royal Institute of Technology.

Fang, N., Zhong, C-Q., Liang, X., Tang, X-F., & Tang, B. (2010). Improvement of extracellular production of a thermophilic subtilase expressed in *Escherichia coli* by random mutagenesis of its N-terminal propeptide. *Appl Microbiol Biotechnol, 85* (5), 1473–1481. doi:10.1007/s00253-009-2183-5

Ferguson, G. P., Tötemeyer, S., MacLean, M. J., Booth, I. R. (1998). Methylglyoxal production in bacteria: suicide or survival? *Arch Microbiol, 170* (4), 209–219.

Ferrer-Miralles, N., Domingo-Espín, J., Corchero, J. L., Vázquez, E., & Villaverde, A. (2009). Microbial factories for recombinant pharmaceuticals. *Microb Cell Fact, 8*: 17. doi:10.1186/1475-2859-8-17

Fiedler, M., & Skerra, A. (2001). *proBA* complementation of an auxotrophic *E. coli* strain improves plasmid stability and expression yield during fermenter production of a recombinant antibody fragment. *Gene, 274* (1-2), 111-118.

Fischer, B. (2013). *Promotoraustausch zum Stressabbau in Wirtszellen und zur Verbesserung der Expressionsrate eines rekombinanten Gens*, Bachelor thesis, Faculty of Technology, Bielefeld University.

Francetic, O., Belin, D., Badaut, C., & Pugsley, A.P. (2000). Expression of the endogenous type II secretion pathway in *Escherichia coli* leads to chitinase secretion. *EMBO J, 19* (24), 6697–6703. doi:10.1093/emboj/19.24.6697

Friehs, K. (2004). Plasmid copy number and plasmid stability. In New Trends and Developments in Biochemical Engineering. *86*, 47-82. Edited by Scheper, T. Springer-Verlag Berlin Heidelberg.

Fu, J., Wilson, D. B., & Shuler, M. L. (1993). Continuous, high level production and excretion of a plasmid-encoded protein by *Escherichia coli* in a two-stage chemostat. *Biotechnol Bioeng, 41*, 937–946.

Gibson, D. G., Young, L., Chuang, R-Y., Venter, J. C., Hutchison, C. A., & Smith, H. O. (2009). Enzymatic assembly of DNA molecules up to several hundred kilobases. *Nat Methods, 6* (5), 343–345. doi:10.1038/NMETH.1318.

Gil, N., Gil, C., Amaral, M. E., Costa, A. P., & Duarte, A. P. (2009). Use of enzymes to improve the refining of a bleached *Eucalyptus globulus* kraft pulp. *Biochem Eng J, 46* (2), 89-95.

Gilbert, P. (1985). The theory and relevance of continuous culture. *J Antimicrob Chemother, 15,* Suppl A, 1–6.

Gilsbach, R., Kouta, M., Bönisch, H., & Brüss, M. (2006). Comparison of in vitro and in vivo reference genes for internal standardization of real-time PCR data. *BioTechniques, 40* (2), 173–177. doi:10.2144/000112052

Glenting, J., Madsen, S. M., Vrang, A., Fomsgaard, A., & Israelsen, H. (2002). A plasmid selection system in *Lactococcus lactis* and its use for gene expression in *L. lactis* and human kidney fibroblasts. *Appl Environ Microbiol, 68* (10), 5051-5056.

Godornes, C., Leader, B. T., Molini, B. J., Centurion-Lara, A., & Lukehart, S. A. (2007). Quantitation of rabbit cytokine mRNA by real-time RT-PCR. *Cytokine, 38* (1), 1–7.

Goh, S., & Good, L. (2008). Plasmid selection in *Escherichia coli* using an endogenous essential gene marker. *BMC Biotechnol*, *8*: 61. doi:10.1186/1472-6750-8-61.

Gottesman, S., Roche, E., Zhou, Y., & Sauer, R. T. (1998). The ClpXP and ClpAP proteases degrade proteins with carboxy-terminal peptide tails added by the SsrA-tagging system. *Genes Dev*, *12*, 1338–1347.

Groeneveld, P., Stouthamer, A. H., & Westerhoff, H. V. (2009). Super life - how and why 'cell selection' leads to the fastest-growing eukaryote. *FEBS J*, *276* (1), 254–270.

Gupta, R., Sharma, P., & Vyas, V. V. (1995). Effect of growth environment on the stability of a recombinant shuttle plasmid, pCPPS-31, in *Escherichia coli*. *J Biotechnol*, *41* (1), 29-37.

Hägg, P., de Pohl, J. W., Abdulkarim, F., & Isaksson, L. A. (2004). A host/plasmid system that is not dependent on antibiotics and antibiotic resistance genes for stable plasmid maintenance in *Escherichia coli*. *J Biotechnol*, *111* (1), 17-30.

Hannig, G., & Makrides, S.C. (1998). Strategies for optimizing heterologous protein expression in *Escherichia coli*. Trends Biotechnol, *16* (2), 54–60.

Harris, P. J., & Fincher, G. B. (2009). Distribution, fine structure and function of (1,3;1,4)-β-Glucans in the grasses and other taxa. In Chemistry, Biochemistry and Biology of (1→3)-β-Glucans and Related Polysaccharides. Ch. 4.6, 621-654. Edited by Bacic, A., Fincher, G. B., & Stone, B. A. Elsevier Academic Press, Burlington, USA.

Harwood, C. R., & Cranenburgh, R. (2008). *Bacillus* protein secretion: an unfolding story. *Trends Microbiol*, *16* (2), 73–79. doi:10.1016/j.tim.2007.12.001

Hayes, A., Zhang, N., Wu, J., Butler, P. R., Hauser, N. C., Hoheisel, J. D., Lim, F. L., Sharrocks, A. D., & Oliver, S. G. (2002). Hybridization array technology coupled with chemostat culture: Tools to interrogate gene expression in *Saccharomyces cerevisiae*. *Methods*, *26* (3), 281–290. doi:10.1016/S1046-2023(02)00032-4

Hellmuth, K., Korz, D. J., Sanders, E. A., & Deckwer, W-D. (1994). Effect of growth rate on stability and gene expression of recombinant plasmids during continuous and high cell density cultivation of *Escherichia coli* TG1. *J Biotechnol*, *32* (3), 289–298.

Herbert, D., Elsworth, R., & Telling, R. C. (1956). The continuous culture of bacteria; a theoretical and experimental study. *J Gen Microbiol*, *14* (3), 601–622.

Hiratsu, K., Shinagawa, H., & Makino, K. (1995). Mode of promoter recognition by the *Escherichia coli* RNA polymerase holoenzyme containing the σ^S subunit: identification of the recognition sequence of the *fic* promoter. *Mol Microbiol*, *18* (5), 841–850.

Hortsch, R., Löser, C., & Bley, T. (2008). A Two-stage CSTR Cascade for Studying the Effect of Inhibitory and Toxic Substances in Bioprocesses. *Eng Life Sci*, *8* (6), 650–657. doi:10.1002/elsc.200800072.

Hoskisson, P. A, & Hobbs, G. (2005). Continuous culture - making a comeback? *Microbiology*, *151* (Pt 10), 3153–3159. doi:10.1099/mic.0.27924-0

Hu, H., & Wood, T. K. (2010). An evolved *Escherichia coli* strain for producing hydrogen and ethanol from glycerol. *Biochem Biophys Res Commun*, *391* (1), 1033–1038.

Huang, C-J., Lin, H., & Yang, X. (2012). Industrial production of recombinant therapeutics in *Escherichia coli* and its recent advancements. *J Ind Microbiol Biotechnol*, *39* (3), 383-399. doi:10.1007/s10295-011-1082-9.

Ihssen, J., & Egli, T. (2004). Specific growth rate and not cell density controls the general stress response in *Escherichia coli*. *Microbiology*, *150* (Pt 6), 1637–1648. doi:10.1099/mic.0.26849-0

Imai, Y., Matsushima, Y., Sugimura, T., & Terada, M. (1991). A simple and rapid method for generating a deletion by PCR. *Nucl Acids Res*, *19* (10), 2785.

Jana, S., & Deb, J. K. (2005). Strategies for efficient production of heterologous proteins in *Escherichia coli*. *Appl Microbiol Biotechnol*, *67* (3), 289-298.

Jensen, P. R., & Hammer, K. (1998). The sequence of spacers between the consensus sequences modulates the strength of prokaryotic promoters. *Appl Environ Microbiol*, *64* (1), 82–87.

Kato, C., Kobayashi, T., Kudo, T., & Horikoshi, K. (1986). Construction of an excretion vector: extracellular production of *Aeromonas* xylanase and *Bacillus* cellulases by *Escherichia coli*. *FEMS Microbiol Lett*, *36*, 31–34.

Kawamukai, M., Matsuda, H., Fujii, W., Utsumi, R., & Komano, T. (1989). Nucleotide sequences of *fic* and *fic-1* genes involved in cell filamentation induced by cyclic AMP in *Escherichia coli*. *J Bacteriol*, *171* (8), 4525–4529.

Kayser, A., Weber, J., Hecht, V., & Rinas, U. (2005). Metabolic flux analysis of *Escherichia coli* in glucose-limited continuous culture. I. Growth-rate-dependent metabolic efficiency at steady state. *Microbiology*, *151* (Pt 3), 693-706.

Kennedy, C., Dixon, R. (1977). The nitrogen fixation cistrons of *Klebsiella pneumoniae*. In Genetic Engineering for Nitrogen Fixation, pp . 51–66. Edited by Hollaender, A., *et al*. Springer US. doi: 10.1007/978-1-4684-0880-5_5

Keseler, I. M., Collado-Vides, J., Santos-Zavaleta, A., Peralta-Gil, M., Gama-Castro, S., Muñiz-Rascado, L., Bonavides-Martinez, C., Paley, S., Krummenacker, M., Altman, T., Kaipa, P., Spaulding, A., Pacheco, J., Latendresse, M., Fulcher, C., Sarker, M., Shearer, A. G., Mackie, A., Paulsen, I., Gunsalus, R. P., Karp, P. D. (2011). EcoCyc: a comprehensive database of *Escherichia coli* biology. *Nucleic Acids Res*, 39: D583-590.

Klumpp, S. (2011). Growth-rate dependence reveals design principles of plasmid copy number control. *PLoS One*, *6*(5), e20403. doi:10.1371/journal.pone.0020403.

Knüttgen, F. (2013). *Vektormodifizierung durch Herausnahme der Antibiotikaresistenzgene*, Bachelor thesis, Faculty of Technology, Bielefeld University.

Kolter, R., Inuzuka, M., & Helinski, D. R. (1978). Trans-complementation-dependent replication of a low molecular weight origin fragment from plasmid R6K. *Cell*, *15* (4), 1199–1208.

Kolter, R., Siegele, D. A., & Tormo, A. (1993). The stationary phase of the bacterial life cycle. *Annu Rev Microbiol*, *47* (1), 855–874.

Korz, D. J., Rinas, U., Hellmuth, K., Sanders, E. A, & Deckwer, W-D. (1995). Simple fed-batch technique for high cell density cultivation of *Escherichia coli*. *J Biotechnol*, *39* (1), 59–65.

Kovárová-Kovar, K., & Egli, T. (1998). Growth kinetics of suspended microbial cells: from single-substrate-controlled growth to mixed-substrate kinetics. *Microbiol Mol Biol Rev, 62* (3), 646–666.

Kroll, J., Steinle, A., Reichelt, R., Ewering, C., & Steinbüchel, A. (2009). Establishment of a novel anabolism-based addiction system with an artificially introduced mevalonate pathway: complete stabilization of plasmids as universal application in white biotechnology. *Metab Eng, 11* (3), 168-177.

Kroll, J., Klinter, S., Schneider, C., Voß, I., & Steinbüchel, A. (2010). Plasmid addiction systems: perspectives and applications in biotechnology. *Microb Biotechnol, 3* (6), 634–657. doi:10.1111/j.1751-7915.2010.00170.x

Kroll, J., Klinter, S., & Steinbüchel, A. (2011). A novel plasmid addiction system for large-scale production of cyanophycin in *Escherichia coli* using mineral salts medium. *Appl Microbiol Biotechnol, 89* (3), 593-604. doi: 10.1007/s00253-010-2899-2.

Laemmli, U.K. (1970). Cleavage of structural proteins during the assembly of the head of bacteriophage T4. *Nature, 227*, 680–685.

Lässig, C. (2009). Bachelor thesis - *Fluoreszierende Proteine*, Faculty of Technology, Bielefeld University.

Landini, P., Egli, T., Wolf, J., & Lacour, S. (2014). sigmaS, a major player in the response to environmental stresses in *Escherichia coli*: role, regulation and mechanisms of promoter recognition. *Environ Microbiol Rep, 6* (1), 1–13. doi:10.1111/1758-2229.12112

Lange, R., & Hengge-Aronis, R. (1994). The cellular concentration of the σ^S subunit of RNA polymerase in *Escherichia coli* is controlled at the levels of transcription, translation, and protein stability. *Genes Dev, 8* (13), 1600–1612.

Lange, R., Fischer, D., & Hengge-Aronis, R. (1995). Identification of transcriptional start sites and the role of ppGpp in the expression of *rpoS*, the structural gene for the σ^S subunit of RNA polymerase in *Escherichia coli*. *J Bacteriol, 177* (16), 4676–4680.

Lara, A. R., & Ramírez, O. T. (2012). Plasmid DNA production for therapeutic applications. In Recombinant Gene Expression, *824*, 271–303. Edited by Lorence, A. Springer Science+Business Media, LLC, New York, USA.

Lee, C., Kim, J., Shin, S. G., & Hwang, S. (2006). Absolute and relative QPCR quantification of plasmid copy number in *Escherichia coli*. *J Biotechnol, 123* (3), 273–280.

Lee, J.M. (1992). *Biochemical Engineering*. New Jersey: Prentice-Hall Inc.

Lendenmann, U., Snozzi, M., & Egli, T. (2000). Growth kinetics of *Escherichia coli* with galactose and several other sugars in carbon-limited chemostat culture. *Can J Microbiol, 46* (1), 72–80.

Leong, D. T., Gupta, A., Bai, H. F., Wan, G., Yoong, L. F., Too, H-P., Chew, F. T., Hutmacher, D. W. (2007). Absolute quantification of gene expression in biomaterials research using real-time PCR. *Biomaterials, 28* (2), 203–210. doi:10.1016/j.biomaterials.2006.09.011.

Lin, W-J., Huang, S-W., & Chou, C. P. (2001). High-level extracellular production of penicillin acylase by genetic engineering of *Escherichia coli*. *J Chem Technol Biotechnol, 76* (10), 1030-1037.

Lin-Chao, S., & Bremer, H. (1986). Effect of the bacterial growth rate on replication control of plasmid pBR322 in *Escherichia coli*. *Mol Gen Genet, 203* (1), 143–149.

Lin-Chao, S., Chen, W-T., & Wong, T-T. (1992). High copy number of the pUC plasmid results from a Rom/Rop-suppressible point mutation in RNA II. *Mol Microbiol, 6* (22), 3385–3393.

Livak, K. J., & Schmittgen, T. D. (2001). Analysis of relative gene expression data using real-time quantitative PCR and the $2^{-\Delta\Delta C_T}$ method. *Methods, 25* (4), 402–408. doi:10.1006/meth.2001.1262

Luirink, J., Duim, B., de Gier, J.W.L., & Oudega, B. (1991). Functioning of the stable signal peptide of the pCloDF13-encoded bacteriocin release protein. *Mol Microbiol, 5* (2), 393–399.

Luke, J., Carnes, A. E., Hodgson, C. P., & Williams, J. A. (2009). Improved antibiotic-free DNA vaccine vectors utilizing a novel RNA based plasmid selection system. *Vaccine, 27* (46), 6454-6459.

Maciąg, A., Peano, C., Pietrelli, A., Egli, T., De Bellis, G., & Landini, P. (2011). In vitro transcription profiling of the σ^S subunit of bacterial RNA polymerase: re-definition of the σ^S regulon and identification of σ^S-specific promoter sequence elements. *Nucleic Acids Res, 39* (13), 5338–5355. doi:10.1093/nar/gkr129

Mairhofer, J., Pfaffenzeller, I., Merz, D., & Grabherr, R. (2008). A novel antibiotic free plasmid selection system: advances in safe and efficient DNA therapy. *Biotechnol J, 3* (1), 83-89.

Marie, C., Vandermeulen, G., Quiviger, M., Richard, M., Préat, V., & Scherman, D. (2010). pFARs, plasmids free of antibiotic resistance markers, display high-level transgene expression in muscle, skin and tumour cells. *J Gene Med, 12*, 323–332. doi:10.1002/jgm.1441.

Martin, G. J. O., Knepper, A., Zhou, B., & Pamment, N. B. (2006). Performance and stability of ethanologenic *Escherichia coli* strain FBR5 during continuous culture on xylose and glucose. *J Ind Microbiol Biotechnol, 33* (10), 834–844.

Mashego, M. R., Rumbold, K., De Mey, M., Vandamme, E., Soetaert, W., & Heijnen, J. J. (2007). Microbial metabolomics: past, present and future methodologies. *Biotechnol Lett, 29* (1), 1–16. doi:10.1007/s10529-006-9218-0

Meinander, N. Q., & Hahn-Hägerdal, B. (1997). Fed-batch xylitol production with two recombinant *Saccharomyces cerevisiae* strains expressing *XYL1* at different levels, using glucose as a cosubstrate: a comparison of production parameters and strain stability. *Biotechnol Bioeng, 54* (4), 391-399.

Mendoza-Vargas, A., Olvera, L., Olvera, M., Grande, R., Vega-Alvarado, L., Taboada, B., Jimenez-Jacinto, V., Salgado, H., Juárez, K., Contreras-Moreira, B., Huerta, A. M., Collado-Vides, J., & Morett, E. (2009). Genome-wide identification of transcription start sites, promoters and transcription factor binding sites in *E. coli*. *PLoS One, 4* (10):e7526. doi:10.1371/journal.pone.0007526.

Mergulhão, F.J.M., Summers, D.K., & Monteiro, G.A. (2005). Recombinant protein secretion in *Escherichia coli, Biotechnol Adv, 23* (3), 177–202.

Metzger, S., Schreiber, G., Aizenman, E., Cashel, M., & Glaser, G. (1989). Characterization of the *relA1* mutation and a comparison of *relA1* with new *relA* null alleles in *Escherichia coli*. *J Biol Chem, 264* (35), 21146–21152.

Miksch, G., & Dobrowolski, P. (1995). Growth phase-dependent induction of stationary-phase promoters of *Escherichia coli* in different gram-negative bacteria. *J Bacteriol, 177* (18), 5374–5378.

Miksch, G., Fiedler, E., Dobrowolski, P., & Friehs, K. (1997a). The *kil* gene of the ColE1 plasmid of *Escherichia coli* controlled by a growth-phase-dependent promoter mediates the secretion of a heterologous periplasmic protein during the stationary phase. *Arch Microbiol, 167* (2), 143–150.

Miksch, G., Neitzel, R., Friehs, K., Fiedler, E., & Flaschel, E. (1997b). Extracellular production of a hybrid β-glucanase from *Bacillus* by *Escherichia coli* under different cultivation conditions in shaking cultures and bioreactors. *Appl Microbiol Biotechnol, 47* (2), 120–126.

Miksch, G., Kleist, S., Friehs, K., & Flaschel, E. (2002). Overexpression of the phytase from *Escherichia coli* and its extracellular production in bioreactors. *Appl Microbiol Biotechnol, 59* (6), 685–694. doi:10.1007/s00253-002-1071-z

Miksch, G., Bettenworth, F., Friehs, K., Flaschel, E., Saalbach, A, & Nattkemper, T. W. (2006). A rapid reporter system using GFP as a reporter protein for identification and screening of synthetic stationary-phase promoters in *Escherichia coli*. *Appl Microbiol Biotechnol, 70* (2), 229–236. doi:10.1007/s00253-005-0060-4

Mnif, B., Vimont, S., Boyd, A., Bourit, E., Picard, B., Branger, C., Denamur, E., & Arlet, G. (2010). Molecular characterization of addiction systems of plasmids encoding extended-spectrum β-lactamases in *Escherichia coli*. *J Antimicrob Chemother, 65* (8), 1599-1603.

Monod, J. (1949). The growth of bacterial cultures. *Annu Rev Microbiol, 3*, 371-394.

Narayanan, N., Khan, M., & Chou, C. P. (2010). Enhancing functional expression of heterologous lipase B in *Escherichia coli* by extracellular secretion. *J Ind Microbiol Biotechnol, 37* (4), 349–361. doi:10.1007/s10295-009-0680-2

Ni, Y., & Chen, R. (2009). Extracellular recombinant protein production from *Escherichia coli*. *Biotechnol Lett, 31* (11), 1661-1670.

Nilsson, J., & Skogman, S. G. (1986). Stabilization of *Escherichia coli* tryptophan-production vectors in continuous cultures: A comparison of three different systems. *Nat Biotechnol, 4*, 901-903.

Nolan, T., Hands, R. E., & Bustin, S. A. (2006). Quantification of mRNA using real-time RT-PCR. *Nat Protoc, 1* (3), 1559–1582. doi:10.1038/nprot.2006.236

Notley, L., & Ferenci, T. (1996). Induction of RpoS-dependent functions in glucose-limited continuous culture: what level of nutrient limitation induces the stationary phase of *Escherichia coli*? *J Bacteriol, 178* (5), 1465-1468.

Novick, A., & Szilard, L. (1950). Experiments with the chemostat on spontaneous mutations of bacteria. *Proc Natl Acad Sci USA, 36* (12), 708–719.

Nurminen, M., Hirvas, L., & Vaara, M. (1997). The outer membrane of lipid A-deficient *Escherichia coli* mutant LH530 has reduced levels of OmpF and leaks periplasmic enzymes. *Microbiology, 143*, 1533–1537.

Park, S., Ryu, D. D. Y., & Kim, J. Y. (1990). Effect of cell growth rate on the performance of a two-stage continuous culture system in a recombinant *Escherichia coli* fermentation. *Biotechnol Bioeng, 36* (5), 493–505.

Perkel, J. M. (2013). Transcriptome Analysis: Microarrays, qPCR and RNA-Seq. *Biocompare*. 21-05-2013. Accessed on 19-06-2014. URL: http://www.biocompare.com/Editorial-Articles/137520-Transcriptome-Analysis-Microarrays-qPCR-and-RNA-Seq/

Peubez, I., Chaudet, N., Mignon, C., Hild, G., Husson, S., Courtois, V., De Luca, K., Speck, D., & Sodoyer, R. (2010). Antibiotic-free selection in *E. coli*: new considerations for optimal design and improved production. *Microb Cell Fact*, *9*: 65.

Planas, A. (2000). Bacterial 1,3-1,4-β-glucanases: structure, function and protein engineering. *Biochim Biophys Acta*, *1543* (2), 361–382.

Popov, M., Petrov, S., Nacheva, G., Ivanov, I., & Reichl, U. (2011). Effects of a recombinant gene expression on ColE1-like plasmid segregation in *Escherichia coli*. *BMC Biotechnol*, *11*: 18.

Porter, R.D., Black, S., Pannuri, S., Carlson, A. (1990). Use of the *Escherichia coli* SSB gene to prevent bioreactor takeover by plasmidless cells. *Nat Biotechnol*, *8*, 47-51.

Reinikainen, P., & Virkajärvi, I. (1989). *Escherichia coli* growth and plasmid copy numbers in continuous cultivations. *Biotechnol Lett*, *11* (4), 225–230.

Riethdorf, S., Ulrich, A., Völker, U., & Hecker, M. (1990). Excretion into the culture medium of a *Bacillus* β-glucanase after overproduction in *Escherichia coli*. *Z Naturforsch C*, *45* (3-4), 240-244.

Rinas, U., & Hoffmann, F. (2004). Selective leakage of host-cell proteins during high-cell-density cultivation of recombinant and non-recombinant *Escherichia coli*. *Biotechnol Prog*, *20* (3), 679–687. doi:10.1021/bp034348k

Robbens, J., Raeymaekers, A., Steidler, L., Fiers, W., & Remaut, E. (1995). Production of soluble and active recombinant murine interleukin-2 in *Escherichia coli* : High level expression, kil-induced release and purification. *Protein Expr Purif*, *6* (4), 481-486.

Rodríguez-Carmona, E., Cano-Garrido, O., Dragosits, M., Maurer, M., Mader, A., Kunert, R., Mattanovich, D., Villaverde, A., & Vázquez, F. (2012). Recombinant Fab expression and secretion in *Escherichia coli* continuous culture at medium cell densities: Influence of temperature. *Process Biochem*, *47* (3), 446-452.

Ruiz, N., Kahne, D., & Silhavy, T. J. (2009). Transport of lipopolysaccharide across the cell envelope: the long road of discovery. *Nat Rev Microbiol*, *7* (9), 677–683. doi:10.1038/nrmicro2184

Scholz, O., Thiel, A, Hillen, W., & Niederweis, M. (2000). Quantitative analysis of gene expression with an improved green fluorescent protein. *Eur J Biochem*, *267* (6), 1565–1570.

Seo, J-H., & Bailey, J. E. (1986). Continuous cultivation of recombinant *Escherichia coli*: Existence of an optimum dilution rate for maximum plasmid and gene product concentration. *Biotechnol Bioeng*, *28* (10), 1590–1594. doi:10.1002/bit.260281018.

Shin, H-D., & Chen, R. R. (2008). Extracellular recombinant protein production from an *Escherichia coli lpp* deletion mutant. *Biotechnol Bioeng*, *101* (6), 1288–1296. doi:10.1002/bit.22013.

Shokri, A., Sandén, A. M., & Larsson, G. (2002). Growth rate-dependent changes in *Escherichia coli* membrane structure and protein leakage. *Appl Microbiol Biotechnol*, *58* (3), 386–392. doi:10.1007/s00253-001-0889-0

Shokri, A., Sandén, A. M., & Larsson, G. (2003). Cell and process design for targeting of recombinant protein into the culture medium of *Escherichia coli*. *Appl Microbiol Biotechnol, 60* (6), 654–664. doi:10.1007/s00253-002-1156-8

Sikorski, R. S., & Hieter, P. (1989). A system of shuttle vectors and yeast host strains designed for efficient manipulation of DNA in *Saccharomyces cerevisiae*. *Genetics, 122*, 19–27.

Simonen, M., & Palva, I. (1993). Protein secretion in *Bacillus* species. *Microbiol Rev, 57* (1), 109–137.

Snijder, H.J., & Dijkstra, B.W. (2000). Bacterial phospholipase A: structure and function of an integral membrane phospholipase. *Biochim Biophys Acta, 1488* (1-2), 91–101.

Sommer, B., Friehs, K., Flaschel, E., Reck, M., Stahl, F., & Scheper, T. (2009). Extracellular production and affinity purification of recombinant proteins with *Escherichia coli* using the versatility of the maltose binding protein. *J Biotechnol, 140* (3-4), 194–202.

Sommer, B., Friehs, K., & Flaschel, E. (2010). Efficient production of extracellular proteins with *Escherichia coli* by means of optimized coexpression of bacteriocin release proteins. *J Biotechnol, 145* (4), 350–358.

Sørensen, H. P., & Mortensen, K. K. (2005). Advanced genetic strategies for recombinant protein expression in *Escherichia coli*. *J Biotechnol, 115* (2), 113-128.

Soubrier, F., Cameron, B., Manse, B., Somarriba, S., Dubertret, C., Jaslin, G., Jung, G., Le Caer, C., Dang, D., Mouvault, J. M., Scherman, D., Mayaux, J. F., & Crouzet, J. (1999). pCOR: a new design of plasmid vectors for nonviral gene therapy. *Gene Ther, 6* (8), 1482-1488.

Spexard, M. (2007). *Prozessoptimierung für β-Glucanase-produzierende Escherichia coli-Stämme*, Diploma thesis, Faculty of Technology, Bielefeld University.

Stephens, M. L., Christensen, C., & Lyberatos, G. (1992). Plasmid stabilization of an *Escherichia coli* culture through cycling. *Biotechnol Prog, 8* (1), 1-4.

Stone, B. A. (2009). Chemistry of β-Glucans. In Chemistry, Biochemistry and Biology of $(1\rightarrow3)$-β-Glucans and Related Polysaccharides. Ch. 2.1, 5-46. Edited by Bacic, A., Fincher, G. B., & Stone, B. A. Elsevier Academic Press, Burlington, USA.

Sunya, S., Delvigne, F., Uribelarrea, J-L., Molina-Jouve, C., & Gorret, N. (2012). Comparison of the transient responses of *Escherichia coli* to a glucose pulse of various intensities. *Appl Microbiol Biotechnol, 95* (4), 1021-1034.

Tanaka, K., Takayanagi, Y., Fujita, N., Ishihama, A., & Takahashi, H. (1993). Heterogeneity of the principal σ factor in *Escherichia coli*: the *rpoS* gene product, σ^{38}, is a second principal σ factor of RNA polymerase in stationary-phase *Escherichia coli*. *Proc Natl Acad Sci USA, 90* (8), 3511-3515.

Teich, A, Meyer, S., Lin, H. Y., Andersson, L., Enfors, S-O., & Neubauer, P. (1999). Growth rate related concentration changes of the starvation response regulators σ^S and ppGpp in glucose-limited fed-batch and continuous cultures of *Escherichia coli*. *Biotechnol Prog, 15* (1), 123-129.

Terpe, K. (2006). Overview of bacterial expression systems for heterologous protein production: from molecular and biochemical fundamentals to commercial systems. *Appl Microbiol Biotechnol, 72* (2), 211-222.

Thisted, T., Nielsen, A. K., & Gerdes, K. (1994). Mechanism of post-segregational killing: translation of Hok, SrnB and Pnd mRNAs of plasmids R1, F and R483 is activated by 3'-end processing. *EMBO J, 13* (8), 1950-1959.

Tyo, K. E. J., Ajikumar, P. K., & Stephanopoulos, G. (2009). Stabilized gene duplication enables long-term selection-free heterologous pathway expression. *Nat Biotechnol, 27* (8), 760–765.

Utsumi, R., Kusafuka, S., Nakayama, T., Tanaka, K., Takayanagi, Y., Takahashi, H., Noda, M., & Kawamukai, M. (1993). Stationary phase-specific expression of the *fic* gene in *Escherichia coli* K-12 is controlled by the *rpoS* gene product (σ^{38}). *FEMS Microbiol Lett, 113* (3), 273-278.

Vaiphei, S. T., Pandey, G., & Mukherjee, K. J. (2009). Kinetic studies of recombinant human interferon-gamma expression in continuous cultures of *E. coli*. *J Ind Microbiol Biotechnol, 36* (12), 1453-1458.

Van der Wal, F. J., Luirink, J., & Oudega, B. (1995a). Bacteriocin release proteins: mode of action, structure, and biotechnological application. *FEMS Microbiol Rev, 17* (4), 381–399.

Van der Wal, F. J., Hagen-Jongman, C. M., Oudega, B. & Luirink, J. (1995b). Optimization of bacteriocin-release-protein-induced protein release by *Escherichia coli*: extracellular production of the periplasmic molecular chaperone FaeE. *Appl Microbiol Biotechnol, 44*, 459-465.

Vandesompele, J., De Preter, K., Pattyn, F., Poppe, B., Van Roy, N., De Paepe, A., & Speleman, F. (2002). Accurate normalization of real-time quantitative RT-PCR data by geometric averaging of multiple internal control genes. *Genome Biol, 3* (7): research0034.1–0034.11.

Velur Selvamani, R. S., Telaar, M., Friehs, K., & Flaschel, E. (2014). Antibiotic-free segregational plasmid stabilization in *Escherichia coli* owing to the knockout of triosephosphate isomerase (*tpiA*). *Microb Cell Fact, 13*: 58. doi:10.1186/1475-2859-13-58.

Vidal, L., Pinsach, J., Striedner, G., Caminal, G., & Ferrer, P. (2008). Development of an antibiotic-free plasmid selection system based on glycine auxotrophy for recombinant protein overproduction in *Escherichia coli*. *J Biotechnol, 134* (1-2), 127-136.

Voss, I., & Steinbüchel, A. (2006). Application of a KDPG-aldolase gene-dependent addiction system for enhanced production of cyanophycin in *Ralstonia eutropha* strain H16. *Metab Eng, 8* (1), 66–78. doi:10.1016/j.ymben.2005.09.003

Warikoo, V., Godawat, R., Brower, K., Jain, S., Cummings, D., Simons, E., Johnson, T., Walther, J., Yu, M., Wright, B., McLarty, J., Karey, K.P., Hwang, C., Zhou, W., Riske, F., Konstantinov, K. (2012). Integrated continuous production of recombinant therapeutic proteins. *Biotechnol Bioeng, 109* (12), 3018–3029.

Werbrouck, H., Botteldoorn, N., Uyttendaele, M., Herman, L., & Van Coillie, E. (2007). Quantification of gene expression of *Listeria monocytogenes* by real-time reverse transcription PCR: optimization, evaluation and pitfalls. *J Microbiol Methods, 69* (2), 306–314. doi:10.1016/j.mimet.2007.01.017.

Wessler, S. R., & Calvo, J. M. (1981). Control of *leu* operon expression in *Escherichia coli* by a transcription attenuation mechanism. *J Mol Biol, 149*, 579–597.

Wick, L. M., Quadroni, M., & Egli, T. (2001). Short- and long-term changes in proteome composition and kinetic properties in a culture of *Escherichia coli* during transition from glucose-excess to glucose-limited growth conditions in continuous culture and vice versa. *Environ Microbiol, 3* (9), 588-599.

Wu, K., & Wood, T. K. (1994). Evaluation of the *hok/sok* killer locus for enhanced plasmid stability. *Biotechnol Bioeng*, *44* (8), 912-921.

Wunderlich, M., Taymaz-Nikerel, H., Gosset, G., Ramírez, O. T., & Lara, A. R. (2014). Effect of growth rate on plasmid DNA production and metabolic performance of engineered *Escherichia coli* strains. *J Biosci Bioeng*, *117* (3), 336–342. doi:10.1016/j.jbiosc.2013.08.007

Yamabhai, M., Emrat, S., Sukasem, S., Pesatcha, P., Jaruseranee, N., & Buranabanyat, B. (2008). Secretion of recombinant *Bacillus* hydrolytic enzymes using *Escherichia coli* expression systems. *J Biotechnol*, *133* (1), 50–57. doi:10.1016/j.jbiotec.2007.09.005

Yang, F., Moss, L. G., & Phillips Jr., G. N. (1996). The molecular structure of green fluorescent protein. *Nat Biotechnol*, *14* (10), 1246–1251.

Yazdani, S. S., & Mukherjee, K. J. (2002). Continuous-culture studies on the stability and expression of recombinant streptokinase in *Escherichia coli*. *Bioproc Biosys Eng*, *24* (6), 341-346.

Yoon, S.H., Kim, S.K., & Kim, J.F. (2010). Secretory production of recombinant proteins in *Escherichia coli*. Recent Pat Biotechnol, *4* (1), 23–29.

Yu, P., & San, K-Y. (1992). Protein release in recombinant *Escherichia coli* using bacteriocin release protein. *Biotechnol Prog*, *8* (1), 25–29.

Zaslaver, A., Bren, A., Ronen, M., Itzkovitz, S., Kikoin, I., Shavit, S., Liebermeister, W., Surette, M. G., & Alon, U. (2006). A comprehensive library of fluorescent transcriptional reporters for *Escherichia coli*. *Nat Methods*, *3* (8), 623–628. doi:10.1038/NMETH895

Zgurskaya, H. I., Keyhan, M., & Matin, A. (1997). The σ^S level in starving *Escherichia coli* cells increases solely as a result of its increased stability, despite decreased synthesis. *Mol Microbiol*, *24* (3), 643–651.

Zipper, H., Brunner, H., Bernhagen, J., & Vitzthum, F. (2004). Investigations on DNA intercalation and surface binding by SYBR Green I, its structure determination and methodological implications. *Nucleic Acids Res*, *32* (12), e103. doi:10.1093/nar/gnh101.

Patents:

Daßler, T., & Wich, G. (2011). *Verfahren zur fermentativen Herstellung von Antikörpern*. European Patent EP1903115 B1. Retrieved from Google Patents.

Kottwitz, B., & Maurer, K. H. (1999). *Glucanasehaltiges Waschmittel*. International Patent WO 1999006516 A1. Retrieved from Google Patents.

Online resources:

Coli Genetic Stock Center (CGSC): *E. coli* Genetic Resources at Yale, Yale University, USA. Online information from http://cgsc.biology.yale.edu/

Life Technologies: Real-Time PCR; Understanding Ct Application Note (2011). Retrieved from https://www.lifetechnologies.com/de/de/home/life-science/pcr/real-time-pcr/qpcr-education/pcr-understanding-ct-application-note.html

Qiagen Resource Center: Critical Factors for Successful Real-Time PCR, Real-Time PCR Brochure (2010). Retrieved from http://www.qiagen.com/knowledge-and-support/resource-center/resource-download.aspx?id=f7efb4f4-fbcf-4b25-9315-c4702414e8d6.

8 Appendix

8.1 Plasmid maps

Reference plasmid p582 (fully sequenced)

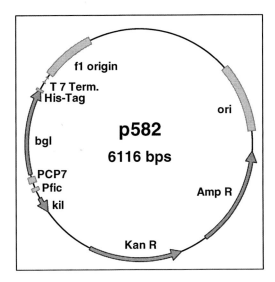

Fig. 8.1: Map of plasmid p582 which represents the starting point for the cloning of auxotrophy complementation constructs described in this work.

Plasmids for GFP fluorescence

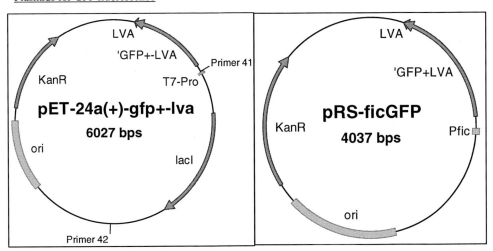

Fig. 8.2: Plasmid maps of initial template pET-24a(+)-gfp+-lva showing binding sites for primers 41 and 42 and the modified pRS-ficGFP containing the *gfp+-lva* reporter gene under the control of the P_{fic} promoter.

pRed/ET expression plasmid

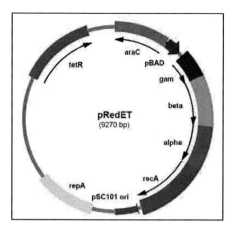

Fig. 8.3: Map of plasmid pRedET (Image taken from Gene Bridges GmbH, Germany). The sequence of pRedET is proprietary. The alternative plasmid supplied in the kit from the manufacturer contained the gene for ampicillin resistance in place of the tetracycline resistance gene (*tetR*) shown here. Single cutters for pRedET (amp) include AhdI, AleI, BaeI, BamHI, BanII, BbsI, EcoNI, KasI, MscI, NarI, NdeI, PciI, PmeI, PvuI, SacI, SalI, SapI, SfoI, SpeI and XmnI among others.

8.2 Sequences

Selection cassette on pFRT

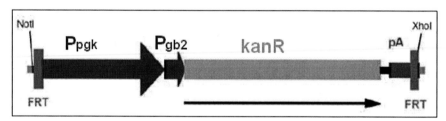

Fig. 8.4: Structure of the selection cassette on pFRT. The cassette is 1637 bp long and contains the kanamycin resistance gene (kanR) under the control of prokaryotic (Pgb2) and eukaryotic (Ppgk) promoters for use in a wide range of target cells. pA is a synthetic polyadenylation sequence for transcriptional termination. The flanking FRT sites allow removal of resistance gene from the genomic target site using FLP-recombinase (Image adapted from Gene Bridges GmbH).

Segment of p582-CP33tpiA

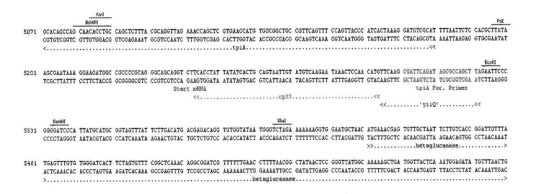

Fig. 8.5: Segment of plasmid p582-CP33tpiA showing the replacement of the original P1/P2 promoter of the complementation gene *tpiA* with the artificial weak promoter CP33. The sequence beginning from the transcription start point has been left unmodified.

bgl gene sequence variation

```
06 Jan 2012                           Alignment Results

Alignment:   Local DNA homologies.
Parameters:  Both strands.  Method:   FastScan - Max Score

  Mol 1 bgl (1 to 720)    Mol 2  p582seq1 (1 to 900)
  Number of sequences to align: 2

  Settings:  Similarity significance value cutoff:  >= 60%

Homology Block:   Percent Matches 99   Score 709   Length 717
                  Mol 1  1 to 717, Mol 2  115 to 831

bgl          1 ATGAAACGAGTGTTGCTAATTCTTGTCACCGGATTGTTTATGAGTTTGTGTGGGATCACTCTTAGTGTTTCGCTCAAAC
p582seq1   115 ATGAAACGAGTGTTGCTAATTCTTGTCACCGGATTGTTTATGAGTTTGTGTGGGATCACTCTTAGTGTTTCGCTCAAAC

bgl         81 AGGCGGATCGTTTTTGAACCTTTAACAGCTATAACTCCGGTTATGGCAAAAGCTGATGTTACTCAAATGGAGATA
p582seq1   195 AGGCGGGATCGTTTTTGAACCTTTAACGGCTATAACTCCGGTTATGGCAAAAGCTGATGTTACTCAAATGGAGATA

bgl        161 TGTTTAACTGACTTGCCGTGCTAATAACGTCTCTATGACGTCATTAGGTGAAATGCGTTGGCGTGACAAGTCCGTCT
p582seq1   275 TGTTTAACTGACTTGGCGTGCGAATAACGTCTCTATGACGTCATGAGGTGAAATGCGTTGGCGTGACAAGTCCGTCT

bgl        241 TATAACAAGTTTGACTGCGGGGAAAACCGCTCGGTTCAAACATATGGCTATGGACTTTATGAAGTCAGAATGAAACCGGC
p582seq1   355 TATAACAAGTTTGACTGCGGGGAAAACCGCTCGGTTCAAACATATGGCTATGGACTTTATGAAGTCAGAATGAAACCGGC

bgl        321 TAAAAACACAGGGATTGTTCATCGTCTTCACTATACAGGTCCAACGGAGGGGACTCCTTGGGATTGAGATTGATATCG
p582seq1   435 TAAAAACACAGGGATTGTTCATCGTCTTCACTATACAGGTCCAACGGAGGGGACTCCTTGGGATTGAGATTGATATCG

bgl        401 AATTTCTAGGAAAGACACGACAAAAGTCAGTTAACTATATACCAATGGGGTTGGCGGTCATGAAAAGGTTATCTCT
p582seq1   515 AATTTCTAGGAAAGACACGACAAAAGTGCAGTTAACTATTATACCAATGGGGTTGGCGGTCATGAAAAGGTTATCTCT

bgl        481 CTTGGCTTTGATGCATCAAAGGGCTTCCATACCTATGCTTTCGATTGGCAGCCAGGGTATATTAAATGGTATGTAGACGG
p582seq1   595 CTTGGCTTTGATGCATCAAAGGGCTTCCATACCTATGCTTTCGATTGGCAGCCAGGGTATATTAAATGGTATGTAGACGG

bgl        561 TGTTTTGAAACATACCGCCACCGCGAATATTCCGAGTACGCCAGGCAAAATTATGATGAATCTATGGAACGAACCGGAG
p582seq1   675 TGTTTTGAAACATACCCCCACCGCGAATATTCCGAGTACGCCAGGCAAAATTATGATGAATCTATGGAACGAACCGGAG

bgl        641 TGGATGACTGGTTAGGTTCTTATAATGGAGCGAATCCGTTGTACGCTGAATATGACTGGGTAAAATATACGAGCAAT
p582seq1   755 TGGATGACTGGTTAGGTTCTTATAATGGAGCGAATCCGTTGTACGCTGAATATGACTGGGTAAAATATACGAGCAAT
```

Fig. 8.6: Sequence variation in the *bgl* gene in p582. bgl refers to the 'typed-in' sequence of β-glucanase gene H1 from Borriss *et al.* (1989). p582seql refers to the sequenced *bgl* gene region from plasmid p582. The base changes with red points above them indicate amino acid changes due to the altered codon.

Fig. 8.7: Region in genome of *E. coli* K12 MG1655 showing the sequences carried as overhangs by primers Up and Low marking the limits for *tpiA* gene deletion. The region marked as 'cloned region' is the extent of *tpiA* sequence cloned into pJET-tpiA.

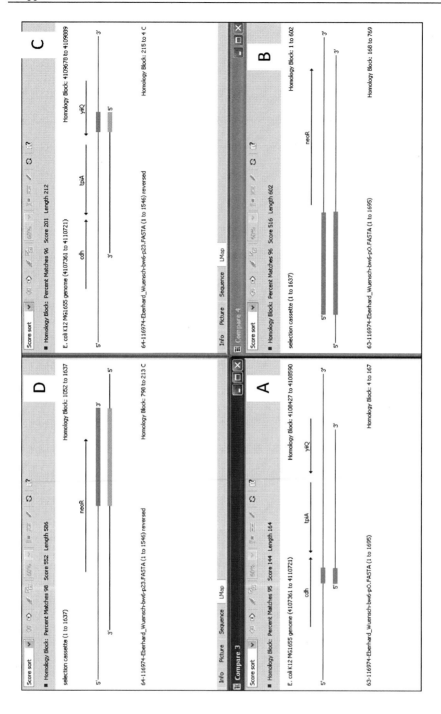

Fig. 8.8: BW6 confirmation - Alignment of sequencing result of 2 kb fragment with genomic region from forward direction with primer O (A) and continuity of base pairs through the selection cassette (B). From the reverse direction using primer 23, alignment of sequencing result with genomic region (C) and continuity through the selection cassette (D). In C and D, discrepancy of 3 base pairs was due to similarity between the genomic region and selection cassette.

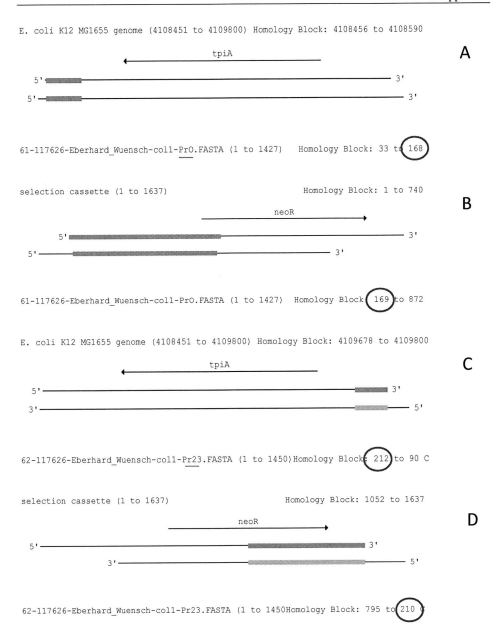

Fig. 8.9: JM1 verfification - Sequencing result of colony 1 shown for representation. Image A shows the sequencing read with primer O (PrO underlined) aligned with the genomic region near the *tpiA* gene on *E. coli*. The alignment stops at position 4108590 on the genome which is exactly the base adjacent to the targeted region. The sequencing read which stops at 168 continues on with 169 being aligned from the first base on the selection cassette shown in image B. Images C and D show the alignment on the other end using the primer 23 (Pr23 underlined). The discrepancy in the 3 bases (210-212) in C and D is due to the fact that they are similar in the genome as well as in the selection cassette.

8.3 Standard calibrations

Table 8.1: Ammonium sulphate calibration

Ammonium sulphate concentration (g L^{-1})	Ammonium ion concentration (g L^{-1})	Parts per million (ppm)	Negative voltage (ΔmV)
0.0367	0.01	10	-40.1
0.367	0.1	100	-94.5
3.67	1	1000	-157.8

Table 8.2: Bradford calibration using Albumin Fraction V

Concentration (mg L^{-1})	OD_{590}	OD_{450}	Quotient (OD_{590}/OD_{450})
0	0.362	0.762	0.475
1	0.382	0.795	0.481
2.5	0.375	0.765	0.490
5	0.37	0.736	0.503
10	0.427	0.757	0.564
25	0.679	0.618	1.099
50	1.03	0.571	1.804
75	1.164	0.509	2.287
100	1.195	0.438	2.728

Table 8.3: Glycerol calibration

Glycerol concentration (g L^{-1})	HPLC Peak area
5	40233
10	84034
20	176567

Publications

Journal publications

1. Velur Selvamani, R. S., Friehs, K., & Flaschel, E. (2014). Extracellular recombinant protein production under continuous culture conditions with *Escherichia coli* using an alternative plasmid selection mechanism. *Bioproc Biosys Eng, 37* (3), 401–413. doi:10.1007/s00449-013-1005-4.

2. Velur Selvamani, R. S., Telaar, M., Friehs, K., & Flaschel, E. (2014). Antibiotic-free segregational plasmid stabilization in *Escherichia coli* owing to the knockout of triosephosphate isomerase (*tpiA*). *Microb Cell Fact, 13*: 58. doi:10.1186/1475-2859-13-58.

Conferences

Lectures

1. Velur Selvamani, R. S., Telaar, M., Friehs, K., & Flaschel, E. (2014). Auxotrophy complementation for antibiotic-free plasmid stabilization. 3rd Minicircle & DNA Vector Conference, Bielefeld.

2. Velur Selvamani, R. S., Friehs, K., & Flaschel, E. (2013). *Kontinuierliche Kultivierung für die Produktion extrazellulärer rekombinanter Proteine in Escherichia coli mittels einer Antibiotika-freien Plasmidselektionsstrategie.* GVC/DECHEMA *Vortrags- und Diskussionstagung*, Bad Wildungen.

Poster presentations

1. Ram Shankar Velur Selvamani, Karl Friehs, Erwin Flaschel (2013). Continuous culture and extracellular recombinant protein production in *Escherichia coli* using alternative plasmid selection mechanism. 2nd European Congress of Applied Biotechnology, The Hague.

2. Velur Selvamani, R. S., Friehs, K., & Flaschel, E. (2013). Continuous culture and extracellular recombinant protein production in *Escherichia coli* using alternative plasmid selection mechanisms. *Frühjahrstagung der Biotechnologen* 2013, Frankfurt.

3. Velur Selvamani, R. S., Friehs, K. & Flaschel, E. (2012). *Kontinuierliche Kultivierung und Produktion extrazellulärer rekombinanter Proteine in Escherichia coli mit alternativen Plasmidselektionsmechanismen.* 30. DECHEMA-*Jahrestagung der Biotechnologen*, Karlsruhe. *Chem Ing Tech, 84*: 1205–1206.

Bisher erschienene Bände der Reihe

Bielefelder Schriften zur molekularen Biotechnologie

ISSN: 2364-4877

Vormals "Bielefelder Schriften zur Zellkulturtechnik"
(ISSN: 1866-9727, Vol. 1-13)

9	Benjamin Müller	Differentielle Analyse des Phosphoproteoms Apoptose-induzierter Jurkat ACC 282-Zellen
		ISBN 978-3-8325-3476-9 38.00 €
10	Sandra Klausing	Optimierung von CHO Produktionszelllinien: RNAi-vermittelter Gen-*knockdown* und Untersuchungen zur Klonstabilität
		ISBN 978-3-8325-3594-0 55.40 €
11	Eva Skerhutt	Proteomanalysen zur Aufklärung regulatorischer Prozesse im Zentralstoffwechsel der humanen Produktionszelllinie AGE1.HN
		ISBN 978-3-8325-3653-4 53.30 €
12	Jennifer Becker-Strugholtz	Transkriptomsequenzierung von CHO-Zelllinien zur Entwicklung eines spezifischen Microarrays für die Analyse des Einflusses wachstumsfördernder Substanzen
		ISBN 978-3-8325-3657-2 49.50 €
13	Christina Timmermann	Transkriptomanalyse von rekombinanten CHO-Zellen. Von der Microarray-Entwicklung bis zur Charakterisierung von Fermentationsprozessen
		ISBN 978-3-8325-3770-8 36.50 €
...	....	....
14	R. S. Velur Selvamani	Continuous culture and extracellular recombinant protein expression in *Escherichia coli*
		ISBN 978-3-8325-3951-1 50.00 €

Alle erschienenen Bücher können unter der angegebenen ISBN im Buchhandel oder direkt beim Logos Verlag Berlin (www.logos-verlag.de, Fax: 030 42 85 10 92) bestellt werden.